MEI STRUCTURED MATHEMATICS

SECOND EDITION

Statistics 2

Anthony Eccles
Nigel Green
Roger Porkess

Series Editor: Roger Porkess

Hodder & Stoughton
A MEMBER OF THE HODDER HEADLINE GROUP

Acknowledgements

We are grateful to the following companies, institutions and individuals who have given permission to reproduce photographs in this book. Every effort has been made to trace and acknowledge ownership of copyright. The publishers will be glad to make suitable arrangements with any copyright holders whom it has not been possible to contact.

Photographs:
Photofusion (page 1); Action-Plus Photographic (pages 3 and 89)

OCR, AQA and Edexcel accept no responsibility whatsoever for the accuracy or method of working in the answers given.

Orders: please contact Bookpoint Ltd, 78 Milton Park, Abingdon, Oxon OX14 4TD. Telephone: (44) 01235 827720, Fax: (44) 01235 400454. Lines are open from 9.00–6.00, Monday to Saturday, with a 24 hour message answering service. Email address: orders@bookpoint.co.uk

British Library Cataloguing in Publication Data
A catalogue record for this title is available from The British Library

ISBN 0 340 771984

First published 1993
Second edition published 2000
Impression number   10 9 8 7 6 5 4 3 2 1
Year   2005 2004 2003 2002 2001 2000

Copyright © 1993, 2000 Anthony Eccles, Nigel Green, Roger Porkess

Typeset by Aarontype Ltd, Easton, Bristol.
Printed in Great Britain for Hodder & Stoughton Educational, a division of Hodder Headline Plc, 338 Euston Road, London NW1 3BH by J. W. Arrowsmith Ltd, Bristol.

MEI Structured Mathematics

Mathematics is not only a beautiful and exciting subject in its own right but also one that underpins many other branches of learning. It is consequently fundamental to the success of a modern economy.

MEI Structured Mathematics is designed to increase substantially the number of people taking the subject post-GCSE, by making it accessible, interesting and relevant to a wide range of students.

It is a credit accumulation scheme based on 45 hour modules which may be taken individually or aggregated to give Advanced Subsidiary (AS) and Advanced GCE (A Level) qualifications in Mathematics, Further Mathematics and related subjects (like Statistics). The modules may also be used to obtain credit towards other types of qualification.

The course is examined by OCR (previously the Oxford and Cambridge Schools Examination Board) with examinations held in January and June each year.

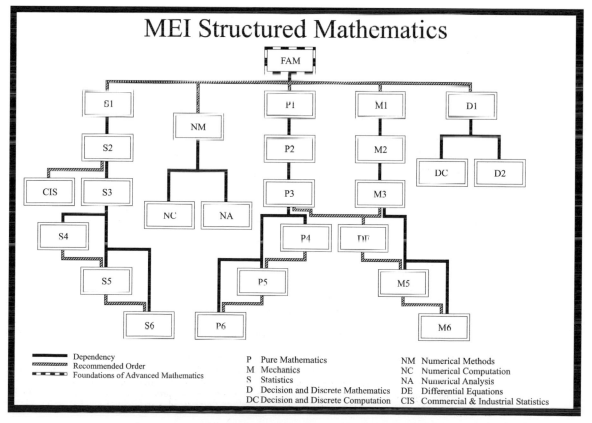

This is one of the series of books written to support the course. Its position within the whole scheme can be seen in the diagram above.

Mathematics in Education and Industry is a curriculum development body which aims to promote the links between Education and Industry in Mathematics at secondary school level, and to produce relevant examination and teaching syllabuses and support material. Since its foundation in the 1960s, MEI has provided syllabuses for GCSE (or O Level), Additional Mathematics and A Level.

For more information about MEI Structured Mathematics or other syllabuses and materials, write to MEI Office, Albion House, Market Place, Westbury, Wiltshire, BA13 3DE.

Introduction

This is the second in a series of books written to support the Statistics modules in MEI Structured Mathematics, but you may also use them for an independent course in the subject. Throughout the course the emphasis is on understanding, interpretation and modelling, rather than on mere routine calculations.

There are four chapters in this book. The techniques for discrete models are covered in the first chapter, and these are illustrated by the Poisson distribution in Chapter 2. In Chapter 3 you are introduced to the normal distribution. The book ends with a chapter on bivariate data, covering correlation and regression lines; you meet two correlation coefficients (Pearson's product moment and Spearman's rank) and use them as statistics for suitable hypothesis tests.

The requirements of the MEI module include a piece of coursework on bivariate data. Detailed guidance on this is available from the MEI office.

Several examples are taken from the pages of a fictional local newspaper, *The Avonford Star*. Much of the information that you receive from the media is of a broadly statistical nature. In these books you are encouraged to recognise this and shown how to evaluate what you are told.

This is the second edition of this book. We would particularly like to draw attention to a change of notation in Chapter 4. In the first edition we use s_{xx}, s_{yy} and s_{xy}, in each case with a divisor n; thus s_{xy} meant $\frac{1}{n} \sum_i (x_i - \bar{x})(y_i - \bar{y})$.

In line with the new specification, in this edition we have adopted the notation S_{xx}, S_{yy} and S_{xy}, with no divisors; thus S_{xy} means $\sum_i (x_i - \bar{x})(y_i - \bar{y})$.

The authors of this book would like to thank the many people who have helped in its preparation and particularly those who read the early versions. We would also like to thank the various examination boards who have given permission for their past questions to be included in the exercises. In addition, thanks are due to Anthony Eccles for his work in preparing the new edition of this book.

Anthony Eccles, Nigel Green and Roger Porkess

Contents

1 Discrete random variables   1

Statistical models   2
Discrete random variables   3
The conditions for a discrete random variable   4
Expectation   10
Expectation of a function X, $E(g[X])$ 14
Expectation algebra   15
Variance   19
Some standard discrete probability distributions   30

2 The Poisson distribution   35

Approximating the binomial terms   37
The Poisson distribution   39
The sum of two or more Poisson distributions   51

3 The normal distribution   59

Using normal distribution tables   61
The normal curve   65
Modelling discrete situations   75
Approximating the binomial distribution   77
Approximating the Poisson distribution   79
The central limit theorem   84
Normal probability graph paper   85

4 Bivariate data

Interpreting scatter diagrams	92
Line of best fit	94
Product moment correlation	96
Interpreting correlation	108
Rank correlation	115
The least squares regression line	125

Appendix 134

1. Mean and variance of the binomial distribution	134
2. Mean and variance of the Poisson distribution	135
3. The sum of two independent Poisson distributions	137
4. Equivalence of Spearman's rank correlation coefficient and Pearson's product moment correlation coefficient	137
5. The least squares regression line	139

Answers 142

Chapter 1	142
Chapter 2	145
Chapter 3	146
Chapter 4	147

Index 152

1 Discrete random variables

An approximate answer to the right problem is worth a good deal more than an exact answer to an approximate problem.

John Tukey

Twins Galore

Out of the 41 babies born in Avonford Maternity Hospital in the last year to parents who live on the Lumumba Estate, no fewer than 6 were twins.

Hospital superintendent Dr Fatima Malik said that while unusual there was nothing particularly significant about this; such multiple-birth clusters are forever occurring somewhere or other she explained.

But self-styled *Guardian of the Environment* Roy James, 55, who lives in Lumumba Drive, says that there must be a source of radioactivity in the area. 'Dr Malik's comment is not the logical explanation', he told me.

Proud brother Ben is delighted with his new sisters, but could radioactivity be causing more multiple births like theirs?

? The newspaper article gives two points of view but how do you decide between them? Before you can make any judgement you must know something about how likely single or multiple births (twins, triplets, etc.) are to occur naturally, that is, their *probability distribution*.

Type of birth	Single baby	Multiple birth
Probability	0.989	0.011

? How does this help you to evaluate the two claims?

There have been 3 pairs of twins born from 38 pregnancies, giving 41 babies in all. The situation can be modelled using the binomial distribution.

Number of pregnancies, $n = 38$.

Probability of a multiple birth, $p = 0.011$.

Probability of a single birth, $q = 0.989$.

Let X be the number of multiple births from this set of pregnancies.

$$P(X = 0) = (0.989)^{38} = 0.656\,839$$
$$P(X = 1) = {}^{38}C_1 \times (0.011)^1 \times (0.989)^{37} = 0.277\,613$$
$$P(X = 2) = {}^{38}C_2 \times (0.011)^2 \times (0.989)^{36} = 0.057\,122$$
$$P(X \geqslant 3) = 1 - P(X \leqslant 2) = 1 - 0.656\,839 - 0.277\,613 - 0.057\,122$$
$$= 0.008\,426$$

As a fraction this result is approximately $\frac{8}{1000} = \frac{1}{125}$. So you would expect a situation like this to occur about once out of every 125 similar community areas throughout the country. Given the large number of such places up and down the country, the doctor's comment that such multiple-birth clusters are always occurring somewhere or other seems perfectly reasonable. Perhaps there is something in what Mr Roy James claimed, however. Certainly if the high multiple birth rates on the Lumumba Estate continue then his claim should be investigated more thoroughly along with other possible causes.

Statistical models

You will find people who will tell you that you can extend the pattern for the probabilities of multiple births as follows:

Number of babies	1	2	3	4	5	. . .
Probability	0.988 89	0.010	0.001	0.0001	0.000 01	. . .

so that the probability for n babies at one pregnancy is given by

$$P = (0.1)^n \qquad\qquad \text{for } n > 1$$
$$\text{and } P = 1 - \sum_{r=2}^{\infty} (0.1)^n = 0.988\,89 \qquad \text{for } n = 1.$$

This is a *mathematical model* to describe the situation. Unfortunately, although algebraically neat, it is a very poor model. The probability of quintuplets is more like one in a hundred million than the one in a hundred thousand suggested by this model.

In statistics you are often looking for models to describe and explain the data you find in the real world. In this chapter you are introduced to some of the techniques for working with models for discrete data, but you should always remember that your results can never be better than your model.

It is always tempting to look for tidy algebraic relationships, like the one proposed for multiple births, but in practice this may just not be possible.

In some cases, like tossing an unbiased coin or rolling a fair die, the model is usually very accurate, but in others this is not so. In Chapters 2 and 3 of this book you will meet two standard distributions, the Poisson and the normal. These, and other theoretical distributions, are often used to model real life situations. The fit can be very good, but you should not forget that they are only being used as models to describe the actual situation. A model which is based on data, like that for the probability of multiple births, can never be more accurate than the actual data, so you should always be prepared to investigate the source of the figures.

The probabilities given for multiple births are derived from national data collected over many years.

Discrete random variables

The number of babies at any pregnancy is an example of a *discrete random variable*. It is *discrete* because it takes only particular values, 1, 2, 3, 4, etc.; you cannot have 2.4 or 0.315 babies. A discrete variable does not, in general, have to be a positive whole number. The shoe sizes of a set of students is a discrete variable taking the possible values ... 5, $5\frac{1}{2}$, 6, $6\frac{1}{2}$, ... By contrast a baby's birth weight is a *continuous* variable which can take any value between two sensible limits. Typical continuous variables are people's heights, weights of elephants and the marathon times of a set of runners.

The number of babies born at any pregnancy is *random* because on becoming pregnant a woman cannot determine how many babies she is going to have. Having twins or triplets is a chance event. (The effects of fertility drugs and in vitro fertilisation have been ignored.)

It is a *variable* because it takes different numerical values.

Notation

A discrete random variable is usually denoted by an upper case letter, such as X, Y, or Z, etc. You may think of this as the name of the variable. The particular values the variable takes are denoted by lower case letters, such as x, y, z. Sometimes these are given suffices x_1, x_2, x_3 etc. Thus $P(X = x_1)$ means 'the probability that discrete random variable X takes the value x_1.'

The conditions for a discrete random variable

If the outcome of a process can be stated as a number, X, which can take different possible values $x_1, x_2, \ldots, x_n$ at random, then X is a discrete random variable. The probabilities $p_1, p_2, \ldots, p_n$ of the different values must sum to 1.

Outcome	x_1	x_2	x_3	$\ldots$	x_n
Probability	p_1	p_2	p_3	$\ldots$	p_n

$$p_1 + p_2 + p_3 + \cdots + p_n = 1 \qquad p_i \geqslant 0, \text{ for } i = 1, \ldots, n$$

Another way of saying this is that the various outcomes cover all possibilities, that is they are *exhaustive*.

Note

Since p_1, p_2 etc. are probabilities, none of them may exceed 1.

This gives the probability distribution of X. The rule which assigns probabilities to the various outcomes is known as the probability function of X. Sometimes it is possible to write the probability distribution as a formula.

Diagrams of probability distributions

If you wish to draw a diagram to show the distribution of a discrete random variable, you must show clearly that the variable is indeed discrete, as in the two diagrams of figure 1.1.

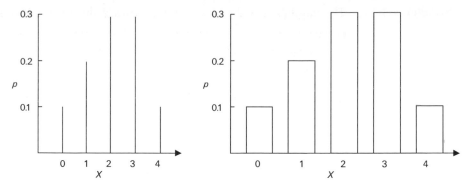

Figure 1.1

This may be contrasted with the distribution of a *continuous* variable which is drawn as a continuous curve, as in figure 1.2. Its equation is called the *probability density function* and usually written f(x). The area under this curve represents probability.

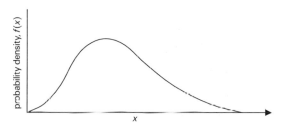

Figure 1.2

You will often find yourself wanting to know the probability that a random variable takes a value no more than a certain number, $P(X \leqslant x)$. This is known as the *cumulative distribution function* and generally denoted by F(x), see figure 1.3.

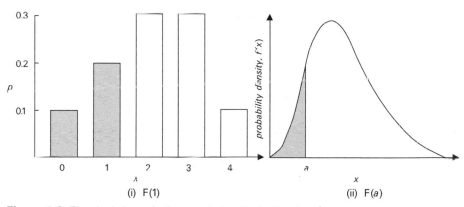

(i) F(1)   (ii) F(a)

Figure 1.3 *The shaded area is the cumulative distribution function*

EXAMPLE 1.1

The variable X is the score on an unbiased die after it is rolled.

(i)   Write down the probability distribution of X.

(ii) Show that X satisfies the conditions for it to be a discrete random variable.

SOLUTION

(i)   The probability distribution of X is:

Outcome	1	2	3	4	5	6
Probability	$\frac{1}{6}$	$\frac{1}{6}$	$\frac{1}{6}$	$\frac{1}{6}$	$\frac{1}{6}$	$\frac{1}{6}$

Notice that $\mathrm{P}(X = 1) + \mathrm{P}(X = 2) + \cdots + \mathrm{P}(X = 6) = 1$

because $\frac{1}{6} + \frac{1}{6} + \cdots + \frac{1}{6} = 1$.

(ii) The set of possible values of X, $\{1, 2, 3, 4, 5, 6\}$, form a discrete set. The outcome from rolling a die is clearly random.

Therefore X is a discrete random variable.

EXAMPLE 1.2

A card is selected at random from a normal pack of 52 playing cards. If it is a spade the experiment ends, otherwise it is replaced, the pack is shuffled and another card is selected. What is the probability distribution of X, the number of cards selected up to and including the first spade?

SOLUTION

This situation is shown in the tree diagram below. You will see that there is no limit to the value which X may take, although larger values of X become more and more unlikely.

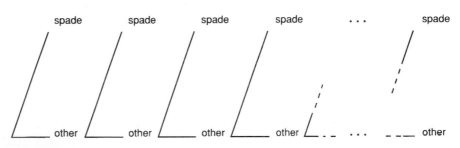

Figure 1.4

x	1	2	3	4	...	r
$\mathrm{P}(X = x)$	$\frac{1}{4}$	$\frac{3}{4} \times \frac{1}{4}$	$\left(\frac{3}{4}\right)^2 \times \frac{1}{4}$	$\left(\frac{3}{4}\right)^3 \times \frac{1}{4}$	...	$\left(\frac{3}{4}\right)^{r-1} \times \frac{1}{4}$

So the probability distribution is given by

$$P(X = r) = \left(\tfrac{3}{4}\right)^{r-1} \times \left(\tfrac{1}{4}\right) \qquad \text{for } r = 1, 2, 3, \ldots$$

Check: Since this is a probability distribution, the sum of all the probabilities should be 1. Is it true that

$$\tfrac{1}{4} + \tfrac{3}{4} \times \tfrac{1}{4} + \left(\tfrac{3}{4}\right)^2 \times \tfrac{1}{4} + \left(\tfrac{3}{4}\right)^3 \times \tfrac{1}{4} + \cdots$$
$$+ \left(\tfrac{3}{4}\right)^{r-1} \times \tfrac{1}{4} + \cdots = 1?$$

The sequence of probabilities form a geometric distribution. You will recognise that it is an infinite geometric series, with the first term, $a = \tfrac{1}{4}$, and the common ratio, $r = \tfrac{3}{4}$.

Substituting these values in the formula for the sum of an infinite geometric series gives:

$$S_\infty = \frac{a}{1-r} = \frac{\tfrac{1}{4}}{1 - \tfrac{3}{4}} = \frac{\tfrac{1}{4}}{\tfrac{1}{4}} = 1 \text{ as required.}$$

EXAMPLE 1.3

The probability distribution of a random variable Y is given by

$$P(Y = y) = cy \qquad \text{for } y = 1, 2, 3, 4, 5.$$

(i) Given that c is a constant find the value of c.
(ii) Hence find the probability that $Y > 3$.

SOLUTION

(i) The relationship $P(Y = y) = cy$ gives the table

Y	1	2	3	4	5
$P(Y)$	c	$2c$	$3c$	$4c$	$5c$

and since Y is a random variable $c + 2c + 3c + 4c + 5c = 1$
$$15c = 1$$
$$c = \tfrac{1}{15}$$

(ii) $P(Y > 3) = P(Y = 4) + P(Y = 5)$
$$= \tfrac{4}{15} + \tfrac{5}{15}$$
$$= \tfrac{9}{15}$$
$$= \tfrac{3}{5}$$

1 The probability distribution of a discrete random variable, X, is given by

$$P(X = x) = kx$$

where k is a constant, for $x = 1, 2, 3, 4$.
Find the value of k.

2 The probability that a variable, Y, takes the value y is given by

$$P(Y = y) = (\tfrac{1}{6})(\tfrac{5}{6})^{y-1} \quad \text{for } y = 1, 2, 3, \ldots$$

Show that Y satisfies the conditions for it to be a discrete random variable and suggest a situation it could be modelling.

3 The probability distribution of a discrete random variable Z is given by

$$P(Z = z) = \frac{az}{8} \quad \text{for } z = 2, 4, 6, 8.$$

(i) Find the value of a.
(ii) Hence find $P(Z < 6)$.

4 The random variable X is given by the number of heads obtained when five fair coins are tossed. Write out the probability distribution for X.

5 The probability distribution of a discrete random variable Y is given by

$$P(Y = y) = cy \quad \text{for } y = 1, 2, 3, 4, 5, 6, 7.$$

where c is a constant.

Find the value of c and the probability that Y is less than 3.

6 The random variable X is given by the sum of the scores when two ordinary dice are thrown.

Write out the probability distribution of X and verify that X is a discrete random variable.

7 The random variable Y is the difference of the scores when two ordinary dice are thrown.
(i) Write out the probability distribution of Y.
(ii) Find $P(Y < 3)$.

8 The random variable Z is the number of heads obtained when four fair coins are tossed.
(i) Write out the probability distribution of Z.
(ii) Find the probability that there are more heads obtained than tails.

9 A box contains six black and four red pens. Three pens are taken at random from the box. The random variable X is the number of red pens obtained. Find the probability distribution of X.

10 Three committee members are to be selected from six men and seven women. Write out the probability distribution for the number of men appointed to the committee assuming the selection is done at random.

11 Two tetrahedral dice each with faces labelled 1, 2, 3 and 4 are thrown and the random variable X is the product of the numbers on which the dice fall.
 (i) Find the probability distribution of X.
 (ii) What is the probability that any throw of the dice results in a value of X which is an odd number?

12 Four cards are drawn, without replacement, from a normal pack of 52. Write the probability distribution for the number of red cards chosen.

13 An ornithologist carries out a study of the numbers of eggs laid per pair by a species of rare bird in its annual breeding season. He concludes that it may be considered as a discrete random variable X with probability distribution given by

$$P(X = 0) = 0.2$$
$$P(X = x) = k(4x - x^2) \quad \text{for } x = 1, 2, 3 \text{ or } 4$$
$$P(X = x) = 0 \qquad\qquad \text{for } x > 4.$$

 (i) Find the value of k and write out the probability distribution as a table.

The ornithologist observes that the probability of survival (that is of an egg hatching and of the chick living to the stage of leaving the nest) is dependent on the number of eggs in the nest. He estimates these probabilities to be as follows.

x	Probability of survival
1	0.8
2	0.6
3	0.4

 (ii) Find, in the form of a table, the probability distribution of the number of chicks surviving per pair of adults.

14 A sociologist is investigating the changing pattern of the numbers of children which women have in a country. She denotes the present number by the random variable X which she finds to have the following distribution.

Number of children x	0	1	2	3	4	5+
Probability $P(X = x)$	0.09	0.22	a	0.19	0.08	negligible

 (i) Find the value of a.

She is anxious to find an algebraic expression for the probability distribution and tries

$$P(X = x) = k(x + 1)(5 - x) \quad \text{for } 0 \leqslant x \leqslant 5$$

$$\text{(x may only take integer values)}$$

$$= 0 \text{ otherwise}$$

(ii) Find the value of k for this model.

(iii) Compare the algebraic model with the probabilities she found. Do you think it is a good model? Is there any reason why it must be possible to express the probability distribution in a neat algebraic form?

15 In a game, each player throws four ordinary six-sided dice. The random variable X is the largest number showing on the dice.

(i) Find the probability that $X = 1$.

(ii) Find the probability that $X \leqslant 2$ and deduce that the probability that $X = 2$ is $\frac{5}{432}$.

(iii) Find the probability that $X = 3$.

(iv) Find the probability that $X = 6$ and explain without further calculation why 6 is the most likely value of X.

[MEI]

Expectation

Figure 1.5

This advertisement was placed by Claire Dadzie who designs and prints T-shirts. She sells them direct to the public for £9 each, avoiding all middlemen.

Claire's T-shirts come in four sizes: small, medium, large and extra large and her profits per sale are £5, £4.50, £4 and £3.50 respectively. She believes that 30% of her orders are for small size, 40% for medium size, 20% for large size and 10% for extra large.

How much profit can she expect to make if she sells 500 T-shirts?

What is her average profit per sale?

Such questions, which are very important to Claire if she is to stay in business, involve the idea of *expectation*.

Claire has a model for the probability distribution of different values of her profit per sale:

Size	Small	Medium	Large	Extra large
Profit (£)	5.00	4.50	4.00	3.50
Probability	0.3	0.4	0.2	0.1

If she sells 500 T-shirts, she expects their sizes to be as follows:

Small	$500 \times 0.3 = 150$
Medium	$500 \times 0.4 = 200$
Large	$500 \times 0.2 = 100$
Extra large	$500 \times 0.1 = \underline{50}$
Total	500

In this case her total profit will be

$$\text{Profit} = 150 \times £5.00 + 200 \times £4.50 + 100 \times £4.00 + 50 \times £3.50$$
$$= £2225$$

Her expected average profit per sale will be $£2225 \div 500 = £4.45$.

You could have found these answers in the reverse order, as follows:

Her expected profit per sale is given by:

$$0.3 \times £5.00 + 0.4 \times £4.50 + 0.2 \times £4.00 + 0.1 \times £3.50 = £4.45$$

Her total expected profit on 500 sales is given by

$$500 \times £4.45 = £2225$$

You will see at once that the two approaches are equivalent to each other. The first involved working out the expected profit for a large number of T-shirts and using that to deduce the profit for one; the second started by finding the profit on one T-shirt and multiplying up to find the profit on a large number. The first step of the second approach may easily be generalised to define expectation.

If a discrete random variable, X, takes possible values $x_1, x_2, x_3, \ldots$ with associated probabilities $p_1, p_2, p_3, \ldots$, the expectation $E(X)$ of X is given by

$$E(X) = \sum_i x_i p_i.$$

Thus in the case of Claire's T-shirts,

$$E(\text{Profit}) = \underset{x_1 p_1}{£5.00 \times 0.3} + \underset{x_2 p_2}{£4.50 \times 0.4} + \underset{x_3 p_3}{£4.00 \times 0.2} + \underset{x_4 p_4}{£3.50 \times 0.1}$$

$$= £4.45$$

Note

The terms *average* and *mean* are often used to convey the same idea as expectation but this may cause confusion.

Expectation is actually the mean of the underlying distribution, the *parent population*. Expectation is denoted by μ (pronounced '*mew*'). The terms average and mean can be applied to either the parent population or a particular sample.

So if one day Claire sells 200 T-shirts at an overall profit of £850, that is a sample of her overall sales over a much longer period. On that one day her *average* or *mean* profit per T-shirt is £850 ÷ 200 = £4.25.

The *expectation* of her profit however is her mean long-term profit per T-shirt, and is £4.45. Claire's expected profit per T-shirt can be thought of as the mean profit of a sample as the sum of the frequencies tends to infinity.

Figures defining the parent population are called *parameters* and usually denoted by Greek letters. The letter μ is used for expectation.

EXAMPLE 1.4

What is the expectation of the score when an unbiased die is rolled once?

SOLUTION

The probability distribution is:

Outcome	1	2	3	4	5	6
Probability	$\frac{1}{6}$	$\frac{1}{6}$	$\frac{1}{6}$	$\frac{1}{6}$	$\frac{1}{6}$	$\frac{1}{6}$

The expected score, $E(X) = \Sigma x_i p_i$

$$= 1 \times \tfrac{1}{6} + 2 \times \tfrac{1}{6} + 3 \times \tfrac{1}{6} + 4 \times \tfrac{1}{6}$$
$$+ 5 \times \tfrac{1}{6} + 6 \times \tfrac{1}{6}$$
$$= \tfrac{21}{6} = 3.5$$

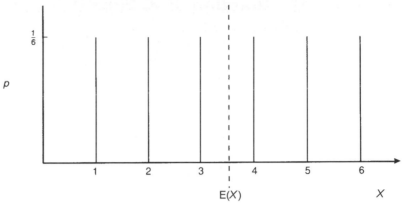

Figure 1.6

Notice that in this case the value of 3.5 for E(X) is an impossible outcome. This is quite often the case (e.g. an average family with 2.4 children) but it is wrong to round it to the nearest apparently sensible value. When used in statistics, expectation is a technical term, with a different meaning from that in everyday English.

EXAMPLE 1.5

A fruit machine is constructed so that the cost is 10p per turn and the amount won by the user in pence at each turn has the following probability distribution:

Payout (pence)	Nil	50	100	500
Less stake	−10	−10	−10	−10
Winnings	−10	40	90	490
Probability	0.94	0.03	0.02	0.01

(i) What is the expected loss per turn?
(ii) How much can the user expect to lose after 20 turns?

SOLUTION

Let the random variable X be the amount won per turn in pence by the user.
(i) $E(X) = -10 \times 0.94 + 40 \times 0.03 + 90 \times 0.02 + 490 \times 0.01$
$= -9.4 + 7.9$
$= -1.5\text{p}$

The expectation of the amount won per turn is −1.5 pence, that is a loss of 1.5 pence.
(ii) After 20 turns the user can expect to lose $20 \times 1.5\text{p} = 30\text{p}$.

Expectation of a function of X, $E(g[X])$

Sometimes you will need to find the expectation of a function of a random variable. That sounds rather forbidding and you may think the same of the definition given below at first sight. However, as you will see in the next two examples, the procedure is straightforward and common sense.

If $g[X]$ is a function of the discrete random variable X then $E(g[X])$ is given by

$$E(g[X]) = \sum_i g[x_i]P(X = x_i)$$

EXAMPLE 1.6

What is the expectation of the square of the number that comes up when a fair die is rolled?

SOLUTION

Let the random variable X be the number that comes up when the die is rolled.

$$g[X] = X^2$$

$$E(g[X]) = E(X^2) = \sum_i x_i^2 P(X = x_i)$$

$$= 1^2 \times \tfrac{1}{6} + 2^2 \times \tfrac{1}{6} + 3^2 \times \tfrac{1}{6} + 4^2 \times \tfrac{1}{6} + 5^2 \times \tfrac{1}{6} + 6^2 \times \tfrac{1}{6}$$

$$= 1 \times \tfrac{1}{6} + 4 \times \tfrac{1}{6} + 9 \times \tfrac{1}{6} + 16 \times \tfrac{1}{6} + 25 \times \tfrac{1}{6} + 36 \times \tfrac{1}{6}$$

$$= \tfrac{91}{6}$$

$$= 15.17$$

Note

This calculation could also have been set out in table form as shown below.

X_i	$P(X=x_i)$	x_i^2	$x_i^2 P(X=x)$
1	$\tfrac{1}{6}$	1	$\tfrac{1}{6}$
2	$\tfrac{1}{6}$	4	$\tfrac{4}{6}$
3	$\tfrac{1}{6}$	9	$\tfrac{9}{6}$
4	$\tfrac{1}{6}$	16	$\tfrac{16}{6}$
5	$\tfrac{1}{6}$	25	$\tfrac{25}{6}$
6	$\tfrac{1}{6}$	36	$\tfrac{36}{6}$
		Σ	$\tfrac{91}{6}$

$$E(g[X]) = \tfrac{91}{6} = 15.17$$

? $E(X^2)$ is not the same as $[E(X)]^2$. In this case $15.57 \neq 3.5^2$ which is 12.25. In fact, the difference between $E(X^2)$ and $[E(X)]^2$ is very important in statistics. Why is this?

EXAMPLE 1.7

A random variable X has the following probability distribution:

Outcome	1	2	3
Probability	0.4	0.4	0.2

(i) Calculate $E(4X + 5)$.

(ii) Calculate $4E(X) + 5$.

(iii) Comment on the relationship between your answers to parts (i) and (ii).

SOLUTION

(i) $E(g[X]) = \sum_i g[x_i] \cdot P(X = x_i)$ with $g[X] = 4X + 5$

x_i	1	2	3
$g[x_i]$	9	13	17
$P(X = x_i)$	0.4	0.4	0.2

$$E(4X + 5) = E(g[X])$$
$$= 9 \times 0.4 + 13 \times 0.4 + 17 \times 0.2$$
$$= 12.2$$

(ii) $E(X) - 1 \times 0.4 + 2 \times 0.4 + 3 \times 0.2 = 1.8$

and so

$$4E(X) + 5 = 4 \times 1.8 + 5$$
$$= 12.2$$

(iii) Clearly $E(4X + 5) = 4E(X) + 5$, both having the value 12.2.

Expectation algebra

In example 1.7 above you found that $E(4X + 5) = 4E(X) + 5$.

The working was numerical, showing that both expressions came out to be 12.2, but it could also have been shown algebraically. This would have been set out as follows:

Proof	Reasons (general rules)
$E(4X + 5) = E(4X) + E(5)$	$E(X \pm Y) = E(X) \pm E(Y)$
$= 4E(X) + E(5)$	$E(aX) = aE(X)$
$= 4E(X) + 5$	$E(c) = c$

Look at the general rules on the right-hand side of the page. (X and Y are random variables, a and c are constants.) They are important but they are also common sense.

Notice the last one, which in this case means the expectation of 5 is 5. Of course it is; 5 cannot be anything else but 5. It is so obvious that sometimes people find it confusing!

These rules can be extended to take in the expectation of the sum of two functions of a random variable.

$$E(f[X] + g[X]) = E(f[X]) + E(g[X])$$

where f and g are both functions of X.

Proof
By definition

$$E(f[X] + g[X]) = \sum_i (f[x_i] + g[x_i]) \cdot P(X = x_i)$$

$$= \sum_i f[x_i] \cdot P(X = x_i) + \sum_i g[x_i] \cdot P(X = x_i)$$

$$= E(f[X]) + E(g[X])$$

EXERCISE 1B

1 Find the expectation of the number of tails when three fair coins are tossed.

2 Find the expectation for the outcome with the following distribution:

Outcome	1	2	3	4	5
Probability	0.2	0.2	0.4	0.1	0.1

3 A discrete random variable X can assume only the values 4 and 5, and has expectation 4.2. Find the two probabilities $P(X=4)$ and $P(X=5)$.

4 A discrete random variable Y can take only the values 50 and 100. Given that $E(Y) = 80$, write out the probability distribution of Y.

5 The probability distribution of a discrete random variable Z is given by

$$P(Z = z) = cz \quad \text{for } z = 2, 3, 4$$

where c is a constant. Find $E(Z)$.

6 A discrete random variable X has the following probability distribution:

x	−2	−1	0	1	2
$P(X=x)$	0.1	0.15	0.15	0.35	0.25

Find
(i) $P(-1 \leqslant X \leqslant 1)$
(ii) $E(X)$
(iii) $P(X < 0.8)$.

7 An unbiased tetrahedral die has faces labelled 2, 4, 6 and 8. If the die lands on the face marked 2, the player has to pay 50p. If it lands on a face marked with a 4 or a 6 the player wins 20p, and if it lands on the face labelled 8 then no money changes hands.

Find the expected gain or loss of the player after
(i) 1 throw
(ii) 3 throws
(iii) 100 throws.

8 Melissa is planning to start a business selling ice-cream. She presents a business plan to her bank manager in which she models the situation as follows.

'Three tenths of days are *warm*, two fifths are *normal* and the rest are *cool*. On warm days I will make £100 (that is the difference between what I get from selling ice-cream and the cost of the ingredients), on normal days £70 and on cool days £50. My additional daily costs will be £45.'
(i) Find Melissa's expected profit per six day week.
(ii) Find Melissa's expected profit after 12 weeks.

9 The discrete random variable X has probability distribution given by

$$P(X = x) = \frac{(4x + 7)}{68} \quad \text{for } x = 1, 2, 3, 4.$$

(i) Find **(a)** $E(X)$ **(b)** $E(X^2)$ **(c)** $E(X^2 + 5X - 2)$.
(ii) Verify that $E(X^2 + 5X - 2) = E(X^2) + 5E(X) - 2$.

10 A discrete random variable X has a probability distribution as follows:

x	0	1	5	20	w
$P(X=x)$	0.1	0.2	0.3	0.1	0.3

Find the value of w in the two cases
(i) $E(X) = 10$
(ii) $E(X) = 12$.

11 Jasmine and Robert are playing a game in which they roll two ordinary dice to obtain a total score, X.
(i) Write down the probability distribution of X.

Halfway through the game their dog eats one of the dice. Robert says 'It's all right. We can play with one die and double its score.' This score is denoted by Y.
(ii) Write down the probability distribution of Y.
(iii) Which of the following are the same and which are different?
  (a) $E(X)$ and $E(Y)$.
  (b) The range of X and the range of Y.
  (c) The mode of X and the mode of Y.

12 A woman has five coins in a box: two £1, two 20p and one 10p. She wants the 10p coin to use in a slot machine and takes the coins out at random, one at a time until she comes to the one she wants. She does not replace a coin once she has taken it out of the box. Find

(i) the expectation of the number of coins she takes out

(ii) the expectation of the amount of money she takes out.

(In both cases include the final 10p.)

13 An experiment consists of throwing two unbiased dice and recording r, the higher of the two numbers shown. When each die shows the same number, this is taken as the value of r. Complete the entries in the following tables, where $p_1, p_2, \ldots$ are the probabilities that $r = 1, 2, \ldots$.

p_1	p_2	p_3	p_4	p_5	p_6
$\frac{1}{36}$	$\frac{3}{36}$				

Verify that the mean of r is $\frac{161}{36}$.

The experiment, as described above, is done *three* times. Find the probabilities that

(i) the three values of r are 3, 4, 5 in any order

(ii) the sum of the three values of r is 16.

[MEI]

14 A radio network is launching a new music game (based on one actually broadcast in France). Each contestant is given the title of a song and a list of seven words, exactly three of which occur in the lyric of the song. The contestant is asked to choose the three correct words (the choice being made before listening to the song). Assuming that a particular contestant makes this choice completely by guesswork, find the probabilities that he gets 0, 1, 2, 3 words correct.

Prizes of £1, £3 and £r are to be awarded for 1, 2 and 3 words correct respectively. Find, in terms of r, the expected prize paid to a contestant choosing completely by guesswork. Hence determine the greatest integer value of r for which the expected prize is less than £3.

[MEI]

15 A company takes on new employees in batches of 80 and immediately gives them training in how to operate and maintain a computer system.

The training course lasts one week, at the end of which the trainees are given a test. If they fail, they repeat the course and take the test again. They are allowed to take the test up to three times but a trainee who has passed the test does not take it again. Those who fail three times are not employed.

In one batch of 80 trainees, the results were as follows:

	Start	Pass	Fail
Week 1	80	32	48
Week 2	48	36	12
Week 3	12	3	9

The company believes this batch to be typical and decides to use these figures to model the costing of its training programme.

(i) Find the probability that somebody passes who is attempting the test at:
(a) the first time **(b)** the second time **(c)** the third time.

(ii) Use your answers to part (i) to deduce the probability that somebody fails all three times and show that the same answer can be obtained by looking at the overall figures.

(iii) Find the probability that somebody who passed the test did so at the second attempt.

It costs the company £4000 overheads + £200 per person for each week that the course is run.

(iv) Find the expected cost per successful trainee.

The company considers, as alternative policies, running the course for only one or two weeks.

(v) Which of the three policies (1, 2 or 3 weeks) would result in the lowest expected cost per trained operator?

Variance

The standard deviation of a discrete random variable gives you a measure of the spread of the distribution about the mean or expected value. The variance is simply the square of the standard deviation.

In *Statistics 1* you learned that

$$\text{standard deviation} = \sqrt{\frac{\sum x^2}{n} - \bar{x}^2}$$

$$\text{and variance} = \frac{\sum x^2}{n} - \bar{x}^2$$

Standard deviation was usually preferred to variance because it is in the same units as those of the data. However, mathematically, it is easier if you now work with the variance. The definition of the variance of a discrete random variable is very similar to that used when finding the variance of a set of numbers.

The variance of a discrete random variable X, $\text{Var}(X)$, is given by

$$\text{Var}(X) = E([X - \mu]^2)$$

Another, and often more convenient, form of this definition is

$$\text{Var}(X) = E(X^2) - \mu^2$$

Proof that the two forms are equivalent:

$$
\begin{aligned}
\text{Var}(X) &= E([X - \mu]^2) \\
&= E(X^2 - 2\mu X + \mu^2) \\
&= E(X^2) - 2\mu E(X) + E(\mu^2) \\
&= E(X^2) - 2\mu^2 + \mu^2 \\
&= E(X^2) - \mu^2
\end{aligned}
$$

This result is sometimes written in the form

$$\text{Var}(X) = E(X^2) - [E(X)]^2$$

$$= \text{expectation of the square} - \text{the square of the expectation}$$

THE AVONFORD STAR

Consumer Concern

Mel Charles, Consumer Affairs correspondent.

Dear Mel,
I am an old age pensioner and my main hobby is gardening. Recently I bought this packet of very expensive seeds from a local company. There were eight seeds in the packet and it said:
'The average germination rate for these seeds is about 50%'.

I followed the planting instructions exactly but not one of the seeds came up. So I went out and bought another packet but exactly the same thing happened. I really want to grow these plants but I certainly can't afford to go on buying dud seeds.
Can you help me?
Yours sincerely,
Reg Harper
(Empty seed packet enclosed)

This is just one of many letters I have received about these seeds which are being sold by The Avonford Seed Co. I went to see their Managing Director, Jill Yates, who was genuinely concerned at the problem and brought in their Research Officer, James Wilkinson, to investigate what had gone wrong.

James set up an experiment. He planted the seeds from a large number of packets and counted the number germinating from each packet. The results surprised him. Given in percentage terms they are:

No. of seeds germinating	% of packets
0	36
1	8
2	4
3	0
4	0
5	1
6	3
7	13
8	35

You can easily work out the average germination rate for the seeds from these figures; it is 51.25%. The claim on the packet was true but it certainly did not tell the whole story.

What is the whole story? It is immediately obvious from the figures that the distribution is very spread out, with most packets having either 0 or 8 seeds germinating. Not one packet had 50% of its seeds germinating.

Clearly you also need to know something about the spread of these figures. The variance is calculated using $\mathrm{Var}(X) = \mathrm{E}(X^2) - \mu^2$. The random variable X is the number of seeds germinating and the percentages found in the experiment are written as probabilities. Since it was a large scale experiment they should be reasonably accurate.

x	0	1	2	3	4	5	6	7	8
$P(X=x)$	0.36	0.08	0.04	0	0	0.01	0.03	0.13	0.35
x^2	0	1	4	9	16	25	36	49	64

The calculation is then as follows:

$$\mathrm{E}(X) = \mu = 0 \times 0.36 + 1 \times 0.08 + 2 \times 0.04 + 3 \times 0 + 4 \times 0$$
$$+ 5 \times 0.01 + 6 \times 0.03 + 7 \times 0.13 + 8 \times 0.35$$
$$= 4.1$$
$$\mathrm{Var}(X) = [0 \times 0.36 + 1 \times 0.08 + 4 \times 0.04 + 9 \times 0$$
$$+ 16 \times 0 + 25 \times 0.01 + 36 \times 0.03 + 49 \times 0.13$$
$$+ 64 \times 0.35] - 4.1^2$$
$$= 13.53$$

So the standard deviation is given by $\sqrt{13.53} = 3.7$ (to one decimal place).

This tells you that a typical value of X is about 3.7 above or below the mean, that is at one extreme of the range or the other. If you knew this you would be warned to expect the distribution to be bimodal, or U-shaped, as shown in figure 1.7.

You would actually expect this to be a binomial distribution, $B(8, 0.5125)$, as shown in figure 1.8. So what has gone wrong?

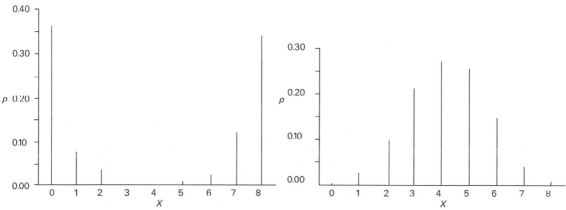

Figure 1.7 *Probability distribution of X, the number of seeds germinating from a packet of 8 seeds*

Figure 1.8 *Probability distribution of B(8, 0.5125)*

The article went on:

So why was the distribution not binomial? The answer is that the trials, in this case whether a seed germinated or not, were not independent. If one seed in a packet did not germinate, it was more likely that the same would be true for the others. Figure 1.7 illustrates a typical bimodal distribution. Such distributions usually arise when the sample is taken from two different populations, as in this case.

EXAMPLE 1.8

The discrete random variable X has the following probability distribution:

x_i	0	1	2	3
$P(X = x_i)$	0.2	0.3	0.4	0.1

Find
(i) $E(X)$
(ii) $E(X^2)$
(iii) $Var(X)$ using (a) $E(X^2) - \mu^2$ (b) $E([X - \mu]^2)$.

SOLUTION

(i) $E(X) = \sum x_i P(X = x_i)$

$\qquad = 0 \times 0.2 + 1 \times 0.3 + 2 \times 0.4 + 3 \times 0.1$

$\qquad = 1.4$

(ii) $E(X^2) = \sum x_i^2 P(X = x_i)$

$\qquad = 0 \times 0.2 + 1 \times 0.3 + 4 \times 0.4 + 9 \times 0.1$

$\qquad = 2.8$

(iii)(a) $Var(X) = E(X^2) - \mu^2$

$\qquad = 2.8 - 1.4^2$

$\qquad = 0.84$

(b) $\text{Var}(X) = E([X - \mu]^2)$

$$= \sum (x_i - \mu)^2 P(X = x_i)$$

$$= (0 - 1.4)^2 \times 0.2 + (1 - 1.4)^2 \times 0.3 + (2 - 1.4)^2 \times 0.4$$

$$+ (3 - 1.4)^2 \times 0.1$$

$$= 0.392 + 0.048 + 0.144 + 0.256$$

$$= 0.84$$

Notice that the two methods of calculating the variance in part (iii) give the same result, as of course they must.

EXAMPLE 1.9   The random variable X has the following probability distribution:

x	1	2	3	4
$P(X=x)$	0.6	0.2	0.1	0.1

Find
(i) $\text{Var}(X)$
(ii) $\text{Var}(7)$
(iii) $\text{Var}(3X)$
(iv) $\text{Var}(3X+7)$.

What general results do answers (ii) to (iv) illustrate?

SOLUTION

(i)

x	1	2	3	4
x^2	1	4	9	16
$P(X=x)$	0.6	0.2	0.1	0.1

$$E(X) = 1 \times 0.6 + 2 \times 0.2 + 3 \times 0.1 + 4 \times 0.1$$

$$= 1.7$$

$$E(X^2) = 1 \times 0.6 + 4 \times 0.2 + 9 \times 0.1 + 16 \times 0.1$$

$$= 3.9$$

$$\text{Var}(X) = E(X^2) - [E(X)]^2$$

$$= 3.9 - 1.7^2$$

$$= 1.01$$

(ii) $\text{Var}(7) = E(7^2) - [E(7)]^2$ *General result*

$\qquad = E(49) - [7]^2$ $\text{Var}(c) = 0$ for a constant c.

$\qquad = 49 - 49$ This result is obvious; a constant is

$\qquad = 0$ constant and so can have no spread.

(iii) $\text{Var}(3X) = E[(3X)^2] - \mu^2$

$= E(9X^2) - [E(3X)]^2$

$= 9E(X^2) - [3E(X)]^2$

$= 9 \times 3.9 - (3 \times 1.7)^2$

$= 35.1 - 26.01$

$= 9.09$

General result

$\text{Var}(aX) = a^2\text{Var}(X)$.

Notice that it is a^2 and not a on the right-hand side, but that taking the square root of each side gives the standard deviation $(aX) = a \times$ standard deviation (X) as you would expect from common sense.

(iv) $\text{Var}(3X + 7)$

$= E[(3X + 7)^2]$

$- [E(3X + 7)]^2$

$= E(9X^2 + 42X + 49)$

$- [3E(X) + 7]^2$

$= E(9X^2) + E(42X) + E(49)$

$- [3 \times 1.7 + 7]^2$

$= 9E(X^2) + 42E(X)$

$+ 49 - 12.1^2$

$= 9 \times 3.9 + 42 \times 1.7$

$+ 49 - 146.41$

$= 9.09$

General result

$\text{Var}(aX + c) = a^2\text{Var}(X)$.

Notice that the constant c does not appear on the right-hand side.

1 The probability distribution of random variable X is as follows:

x	1	2	3	4	5
$P(X = x)$	0.1	0.2	0.3	0.3	0.1

(i) Find (a) $E(X)$ (b) $\text{Var}(X)$.

(ii) Verify that $\text{Var}(2X) = 4\text{Var}(X)$.

2 The probability distribution of a random variable X is as follows:

x	0	1	2
$P(X = x)$	0.5	0.3	0.2

(i) Find (a) $E(X)$ (b) $\text{Var}(X)$.

(ii) Verify that $\text{Var}(5X + 2) = 25\text{Var}(X)$.

3 Prove that $\text{Var}(aX - b) = a^2\,\text{Var}(X)$ where a and b are constants.

4 A coin is biased so that the probability of obtaining a tail is 0.75. The coin is tossed four times and the random variable X is the number of tails obtained. Find

(i) $E(2X)$

(ii) $\text{Var}(3X)$.

5 Birds of a particular species lay either 0, 1, 2 or 3 eggs in their nests, with probabilities as shown in the following table.

Number of eggs	0	1	2	3
Probability	0.25	0.35	0.30	k

Find

(i) the value of k

(ii) the expected number of eggs laid in a nest

(iii) the standard deviation of the number of eggs laid in a nest.

[Cambridge]

6 A committee of three is to be selected at random from three men and four women. The number of men on the committee is the random variable Y. Find

(i) $E(Y)$

(ii) $Var(Y)$.

7 A discrete random variable W has the distribution:

w	1	2	3	4	5	6
$P(W=w)$	0.1	0.2	0.1	0.2	0.1	0.3

Find the mean and variance of

(i) $W+7$

(ii) $6W-5$.

8 The random variable X is the number of heads obtained when four unbiased coins are tossed. Construct the probability distribution for X and find

(i) $E(X)$

(ii) $Var(X)$

(iii) $Var(3X+4)$.

9 A shop sells red and blue refills for pens, and keeps them all in the same box. The box contains five red refills at £2.00 each and four blue ones at £1.60 each. A customer takes three refills out at random, not realising there are different colours in the box. Find the expectation and variance of the amount of money the customer spends.

10 A board game is played by moving a counter S squares forwards at a time, where S is determined by the following rule.

A fair six sided die is thrown once. S is half the number shown on the die if that number is even; otherwise S is twice the number shown on the die.

Write out a table showing the possible values of S and their probabilities. Use your table to calculate the mean and variance of S.

[Cambridge]

11 In an arcade game, whenever a lever is pulled a number appears on a screen. The number can be 1, 2 or 3 and the probabilities of obtaining these numbers are given in the following table.

Number	1	2	3
Probability	0.1	0.5	0.4

(i) The score, S_1, when the lever is pulled for the first time is the number obtained. Calculate the mean and variance of S_1.

(ii) A second score, S_2, is obtained as follows. If $S_1 = 1$ the lever is pulled a second time, and S_2 is the number which then appears on the screen. If $S_1 = 2$ or $S_1 = 3$, the lever is not pulled a second time and S_2 is 0. Tabulate the possible values of $S_1 + S_2$ and their associated probabilities. Hence calculate the mean of $S_1 + S_2$.

[Cambridge]

12 An electronic device produces an output of 0, 1 or 3 volts, with probabilities $\frac{1}{2}$, $\frac{1}{3}$ and $\frac{1}{6}$ respectively. The random variable X denotes the result of adding the outputs for two such devices, which act independently.

(i) Show that $P(X = 4) = \frac{1}{9}$.

(ii) Tabulate all the possible values of X with their corresponding probabilities.

(iii) Hence calculate $E(X)$ and $Var(X)$, giving your answers as fractions in their lowest terms.

[Cambridge]

13 A box contains nine numbered balls. Three balls are numbered 3, four balls are numbered 4 and two balls are numbered 5.

Each trial of an experiment consists of drawing two balls without replacement and recording the sum of the numbers on them, which is denoted by X. Show that the probability that $X = 10$ is $\frac{1}{36}$, and find the probabilities of all other possible values of X.

Use your results to show that the mean of X is $\frac{70}{9}$, and find the standard deviation of X.

Two trials are made. (The two balls in the first trial are replaced in the box before the second trial.) Find the probability that the second value of X is greater than or equal to the first value of X.

[MEI]

14 A curiously shaped six-faced die produces scores, X, for which the probability distribution is given in the following table.

r	1	2	3	4	5	6
$P(X = r)$	k	$\frac{k}{2}$	$\frac{k}{3}$	$\frac{k}{4}$	$\frac{k}{5}$	$\frac{k}{6}$

Show that the constant k is $\frac{20}{49}$, and find the mean and variance of X.

Show that, when this die is thrown twice, the probability of obtaining two equal scores is very nearly $\frac{1}{4}$.

[MEI]

15 A bag contains four balls, numbered 2, 4, 6, 8 but identical in all other respects. One ball is chosen at random and the number on it is denoted by N, so that $P(N = 2) = P(N = 4) = P(N = 6) = P(N = 8) = \frac{1}{4}$. Show that $\mu = E(N) = 5$ and $\sigma^2 = \mathrm{Var}(N) = 5$.

Two balls are chosen at random one after the other, with the first ball being replaced after it has been drawn. Let $\overline{N}$ be the arithmetic mean of the numbers on the two balls. List the possible values of $>\overline{N}$ and their probabilities of being obtained. Hence evaluate $E(\overline{N})$ and $\mathrm{Var}(\overline{N})$.

[MEI]

16 A random number generator in a computer game produces values which can be modelled by the discrete random variable X with probability distribution given by

$$P(X = r) = kr! \quad r = 0, 1, 2, 3, 4$$

where k is a constant.

(i) Show that $k - \frac{1}{34}$ and illustrate the probability distribution with a sketch.

(ii) Find the expectation and variance of X.

Two independent values of X are generated. Let these values be X_1 and X_2.

(iii) Show that $P(X_1 = X_2)$ is a little greater than 0.5.

(iv) Given that $X_1 = X_2$, find the probability that X_1 and X_2 are each equal to 4.

[MEI]

17 A traffic surveyor is investigating the lengths of queues at a particular set of traffic lights during the daytime, but outside rush hours. He counts the number of cars, X, stopped and waiting when the lights turn green on 90 different occasions, with the following results:

No. of cars, x_i	0	1	2	3	4	5	6	7	8	9
Frequency, f_i	3	9	12	15	15	15	11	8	2	0

(i) Use these figures to estimate the probability distribution of the number of cars waiting when the lights turn green.

(ii) Use your probability distribution to estimate the expectation and variance of X.

A colleague of the surveyor suggests that the probability distribution might be modelled by the expression $P(X = x_i) = kx_i(8 - x_i)$.

(iii) Find the value of k.

(iv) Find the values of the expectation and variance of X given by this model.

(v) Do you think it is a good model?

18 A wine bar sells wine by the glass and no other drinks during lunch time. It is usually very busy. In order to help his planning, the owner commissions a mathematics student to do some research into the drinking habits of his customers. After observing a large number of people over a representative period, the student tells the owner '*The probability distribution of* X, *the number of drinks a customer will have, is given by*

$$P(X = x_i) = \frac{(7 - x_i)}{21} \quad \textit{for } x_i = 1, 2, \dots, 6$$

$$\textit{otherwise } 0.'$$

The owner tells the student that he wants the information in a form he can understand.

(i) Write the information in a form that the owner can indeed understand.

(ii) Show that the probability distribution satisfies the conditions for X to be a discrete random variable.

(iii) Calculate the expectation and variance of X.

(iv) Explain the significance to the bar's owner of whether the expectation and variance have large or small values.

The owner considers a possible new sales policy of selling only double glasses of wine (at double the price). He considers two possible models of how this might affect his custom.

(a) The same number of people come to the bar as before.

Those who previously would have had 2, 4 or 6 glasses will now have 1, 2 or 3 double glasses, respectively.

Those who would have had 1, 3 or 5 glasses round up their drinking by having 1, 2 or 3 double glasses, respectively.

(b) Those who previously would have had 2, 4 or 6 glasses will now have 1, 2 or 3 double glasses, respectively.

Half of those who would have had 1, 3 or 5 glasses round up their drinking by having 1, 2 or 3 double glasses, respectively; the other half round down by having 0, 1 or 2 double glasses.

Those who have 0 glasses no longer come to the bar at all.

(v) Calculate the mean and variance for both models.

(vi) Explain how you would interpret the answers to part (v) in terms of running the bar, and state what you would do if you were the owner.

19 I was asked recently to analyse the number of goals scored per game by our local Ladies Hockey Team. Having studied the results for the whole of last season, I proposed the following model, where the discrete random variable X represents the number of goals scored per game by the team:

$$P(X = r) = k(r + 1)(5 - r)^2 \quad \text{for } r = 0, 1, 2, 3.$$

(i) Show that k is 0.01 and illustrate the probability distribution with a sketch.

(ii) Find the expectation and variance of X.

(iii) Assuming that the model is valid for the forthcoming season, find the probability that

    (a) the team will fail to score in the first two games

    (b) the team will score a total of four goals in the first two games.

    What other assumption is necessary to obtain these answers?

(iv) Give two distinct reasons why the model might not be valid for the forthcoming season.

[MEI]

20 Jane removed the labels from four cans of rhubarb and three cans of tuna fish, for use in a competition. She forgot to mark the cans which, without their labels, look identical. She now opens successive cans, chosen at random, looking for a can of tuna fish. If X represents the number of cans she has to open until she finds a can of tuna fish, then you are given that

$$P(X = r) = k(6 - r)(7 - r) \qquad \text{for } r = 1, 2, 3, 4, 5.$$

(i) Copy and complete the table below. Show that $k = \frac{1}{70}$.

r	1	2	3	4	5
$P(X=r)$		$20k$		$6k$	

(ii) Draw a sketch to illustrate the probability distribution. State the modal value and whether the distribution is positively or negatively skewed.

(iii) Find the mean and standard deviation of X.

(iv) A can of rhubarb costs 50p and a can of tuna costs £1.20. Find the expected total cost of the cans Jane has to open to find a can of tuna fish.

(v) Use a probability argument to show that the given formula is correct when $r = 3$.

[MEI]

INVESTIGATIONS

Some of the work in the previous pages may have seemed quite theoretical to you. How does it apply to modelling real life situations?

Here are two practical investigations to help you to answer this for yourself. Make sure that you try at least one of them. It is important that you not only know the theory but also have experience of how it works in practice.

CHARITY STALL

Invent a game for a charity stall where members of the public pay 20 pence a turn to select a card at random from a normal pack of 52. Decide for which outcomes the player is to win and what the winnings should be. Arrange for your game to make an expected profit of 10 pence per turn for the organiser.

As you play the game more and more times, keep a record of the results. You would expect the organiser's average profit per turn to get progressively closer to 10 pence. Does this really happen?

ROLLING A DIE

Work out the mean, E(X), and variance, Var(X), of the score, X, when a single die is rolled.

Roll a die six times and work out the values of the mean and variance of the scores you have obtained. How do they compare with the theoretical values you calculated?

Repeat the experiment but this time roll the die 60 times. Are your answers closer now?

Organise your friends so that, between you, the die is rolled 600 times. Are your answers even closer?

Some standard discrete probability distributions

There are several standard discrete probability distributions which you will often need when modelling statistical data.

The binomial distribution

You will recall that a binomial distribution results from n independent trials of an experiment which has exactly two possible outcomes, often referred to as *success* (with probability p) and *failure* (with probability $q = 1 - p$). The probabilities p and q remain the same from one trial to the next, as for example, in rolling a die or tossing a coin.

> For the binomial distribution, B(n, p):
> Mean $= np$ and Variance $= npq = np(1 - p)$.

(The proofs of these results are given in the Appendix on pages 134–135)

EXAMPLE 1.10

Find the mean and variance of the discrete random variable $X \sim B(8, 0.2)$. (Notice the use of the symbol $\sim$ to mean 'is distributed as'.)

SOLUTION

$$E(X) = np = 8 \times 0.2 = 1.6$$
$$Var(X) = npq = 8 \times 0.2 \times 0.8 = 1.28$$

The uniform distribution

This distribution occurs when all the possible outcomes are equally likely, as for example when a single die is rolled. There are six possible outcomes, each with the same probability, $\frac{1}{6}$, as shown in figure 1.9.

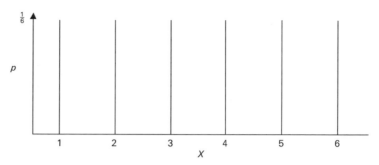

Figure 1.9 *Probability distribution of the score when a single die is rolled*

The mean and variance of this distribution are the same as those of the particular values the variable can take.

The geometric distribution

If a sequence of independent trials is conducted, for each of which the probability of success is p and that of failure q ($q = 1 - p$), and the random variable X is the number of the trial on which the first success occurs, then X has a *geometric distribution*.

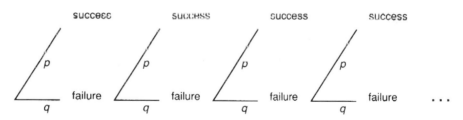

Figure 1.10

x	1	2	3	4	...
Probability	p	pq	pq^2	pq^3	...

Thus the distribution is given by:

$$P(X = x) = pq^{x-1} \qquad \text{for } x \geqslant 1$$

An example of this is the probability distribution of the number of times you have to roll a die until the number six shows.

For a geometric distribution: Mean $= \frac{1}{p}$ and Variance $= \frac{q}{p^2}$.

1 The random variable, R, is the number of heads when a fair coin is tossed four times.

   (i) Copy and complete the table giving the probability distribution of R.

r	0	1	2	3	4
$P(R=r)$	$\frac{1}{16}$	$\frac{4}{16}$			

   (ii) Use these figures to find $E(R)$ and $Var(R)$.

   (iii) Show that you get the same values by using $E(R) = np$ and $Var(R) = npq$.

2 The random variable, X, is the number of sixes obtained when three fair dice are rolled.

   (i) Copy and complete the table giving the probability distribution of X.

x	0	1	2	3
$P(X=x)$	0.5787	0.3472		

   (ii) Use these figures to find $E(X)$ and $Var(X)$.

   (iii) Show that you get the same values by using $E(X) = np$ and $Var(X) = npq$.

3 An applicant for a post in a casino is asked to demonstrate that she can toss a coin fairly. She is to carry out 20 sets of trials, each consisting of 10 tosses of the coin. In each the number of heads is recorded.

   (i) What are the expectation and variance of the numbers of heads in a set of 10 tosses?

   In her 20 sets of trials, the applicant gets the following numbers of heads:

   3 5 4 8 7   7 5 5 5 4   3 4 4 6 6   7 5 6 4 5

   (ii) Calculate the mean and variance of these data.

   (iii) Do you think the applicant is tossing the coin fairly?

4 At a charity fund-raising dinner, each guest is to be asked to roll a fair die and give £5 for each spot on the uppermost face of the die. The random variable Y is the amount raised per person in this way.

   (i) Write out the probability distribution of Y.

   (ii) Find the expectation and the standard deviation of Y.

   There are 100 guests at the dinner. When the money has been collected from the die rolling it is found to amount to £2750.

   (iii) What do you think has happened?

5 A couple decide they will have children until they have a daughter and then stop. Assume that each child is equally likely to be a girl or a boy.

   (i) Find the expectation of the number of children they will have.

   Their friends decide to have children until they have one of each sex.

   (ii) What is the expectation of their number of children?

   (iii) Find, for each couple, the probability that they will end up with at least four children.

6 A box contains three red balls and two green balls. A ball is drawn out at random and then replaced until a green ball is obtained. If X represents the number of draws required to pick a green ball,
(i) describe the probability distribution of X
(ii) find $E(X)$ and $Var(X)$.

7 In a promotion campaign a breakfast cereal company includes between one and four vouchers in each of its packets, the number, X, being random with all four possibilities equally likely.
(i) Write down the probability distribution of X and find $E(X)$ and $Var(X)$.

A family buy two packets of the cereal and receive Y vouchers.
(ii) Write down the probability distribution of Y and find $E(Y)$ and $Var(Y)$.

The promotion lasts for only a few weeks and anybody who gets six or more vouchers is entitled to exchange them for a free gift.
(iii) What is the least number of packets that somebody must buy in order to have a probability of 0.5 or more of qualifying for the gift?
(iv) Mr MacTaggart buys three packets. What is the expectation of the number of gifts he will receive?

8 A coin is biased in such a way that, on any throw, $P(head) = p$ and $P(tail) = q$, where $p + q = 1$. The random variable X denotes the number of heads resulting from three throws of this coin. Tabulate the probability distribution of X and show from first principles (i.e. without quoting any results relating to the binomial distribution) that $E(X) = 3p$ and that $Var(X) = 3pq$.

In an experiment, three throws of the coin were repeated 1000 times. The numbers of times each value of X occurred are shown in the table.

Number of heads	0	1	2	3
Frequency	90	329	412	169

Calculate the mean number of heads from these figures and use it to estimate the value of p.

Hence estimate the probability that five throws of this coin will result in at least one head.

[Cambridge]

9 Dan's hobby is archery. When shooting at a target, he scores 1, 2 or 3 points when he hits the target, depending on how close he is to the centre. If he misses the target completely he scores 0 points. From past experience, the distribution of X, his score for each shot, is as follows:

x	0	1	2	3
$P(X=x)$	0.1	0.4	0.3	0.2

(i)   Tabulate the cumulative distribution function for X, i.e. $P(X \leqslant r)$ for $r = 0, 1, 2, 3$.

Dan has three shots at the target and his scores are independent. Let L be the discrete random variable which represents the largest of the three scores.

(ii) Explain why $P(L \leqslant 2) = [P(X \leqslant 2)]^3 = 0.512$.
Find similarly $P(L \leqslant 1)$.
Write down $P(L = 0)$ and $P(L \leqslant 3)$.

(iii) Use the probabilities calculated in part (ii) to write down the probability distribution for L.

(iv) Find the mean and variance of L.

[MEI]

10 In each round of a general knowledge quiz a contestant is asked up to three questions. The round stops when the contestant gets a question wrong or has answered all three questions correctly. One point is awarded for each correct answer. A contestant who answers all three questions correctly receives in addition one bonus point.

Jayne's probability of getting any particular question correct is 0.7, independently of other questions. Let X represent the number of points she scores in a round.

(i)   Show that $P(X = 1) = 0.21$ and find $P(X = r)$ for $r = 0, 2$ and 4.

(ii) Draw a sketch to illustrate this discrete probability distribution.

(iii) Find the mean and standard deviation of X.

(iv) Find the probability that Jayne obtains a higher score in the second round than she does in the first round.

[MEI]

KEY POINTS

1 For a discrete random variable, X, which can assume only the values $x_1, x_2, \ldots, x_n$ with probabilities $p_1, p_2, \ldots, p_n$ respectively:

- $\sum p_i = 1 \quad p_i \geqslant 0$
- $E(X) = \sum x_i p_i = \sum x_i P(X = x_i)$
- $E(g[X]) = \sum g[x_i] p_i = \sum g[x_i] P(X = x_i)$
- $\text{Var}(X) = E(X^2) - [E(X)]^2$

2 For any discrete random variable X and constants a and c:

- $E(c) = c$
- $E(aX) = aE(X)$
- $E(aX + c) = aE(X) + c$
- $E(f[X] + g[X] = E(f[X]) + E(g[X])$
- $\text{Var}(c) = 0$
- $\text{Var}(aX) = a^2 \text{Var}(X)$
- $\text{Var}(aX + c) = a^2 \text{Var}(X)$

2

The Poisson distribution

If something can go wrong, sooner or later it will go wrong.

Murphy's Law

Rare Disease Blights Town

Chemical Plant Blamed

A rare disease is attacking residents of Avonford. In the last year alone five people have been diagnosed as suffering from it. This is over three times the national average.

The disease (known as *Palfrey's condition*) causes nausea and fatigue. One sufferer, James Louth (32), of Harpers Lane, has been unable to work for the past six months. His wife Muriel (29) said 'I am worried sick, James has lost his job and I am frightened that the children (Mark, 4, and Samantha, 2) will catch it.'

Mrs Louth blames the chemical complex on the industrial estate for the disease. 'There were never any cases before *Avonford Chemicals* arrived.'

Local environmental campaigner Roy James supports Mrs Louth. 'I warned the local council when planning permission was sought that this would mean an increase in this sort of illness. Normally we would expect 1 case in every 40 000 of the population in a year.'

Avonford Chemicals spokesperson, Julia Millward said, 'We categorically deny that our plant is responsible for the disease. Our record

Muriel Louth believes the local chemical plant could destroy her family's lives

on safety is very good. None of our staff has had the disease. In any case five cases in a population of 60 000 can hardly be called significant.'

The expected number of cases is $60\,000 \times \frac{1}{40\,000}$ or 1.5, so 5 does seem rather high. Do you think that the chemical plant is to blame or do you think people are just looking for an excuse to attack it? How do you decide between the two points of view? Is 5 really that large a number of cases anyway?

The situation could be modelled by the binomial distribution. The probability of somebody getting the disease in any year is $\frac{1}{40\,000}$ and so that of not getting it is $1 - \frac{1}{40\,000} = \frac{39\,999}{40\,000}$.

The probability of 5 cases among 60 000 people (and so 59 995 people not getting the disease) is given by

$$^{60\,000}C_5 \left(\frac{39\,999}{40\,000}\right)^{59\,995} \left(\frac{1}{40\,000}\right)^5 \approx 0.0141$$

What you really want to know, however, is not the probability of exactly 5 cases but that of 5 or more cases. If that is very small, then perhaps something unusual did happen in Avonford last year.

You can find the probability of 5 or more cases by finding the probability of up to and including 4 cases, and subtracting it from 1.

The probability of up to and including 4 cases is given by:

$$\left(\frac{39\,999}{40\,000}\right)^{60\,000} \qquad\qquad \text{0 cases}$$

$$+\,^{60\,000}C_1 \left(\frac{39\,999}{40\,000}\right)^{59\,999} \left(\frac{1}{40\,000}\right) \qquad\qquad \text{1 case}$$

$$+\,^{60\,000}C_2 \left(\frac{39\,999}{40\,000}\right)^{59\,998} \left(\frac{1}{40\,000}\right)^2 \qquad\qquad \text{2 cases}$$

$$+\,^{60\,000}C_3 \left(\frac{39\,999}{40\,000}\right)^{59\,997} \left(\frac{1}{40\,000}\right)^3 \qquad\qquad \text{3 cases}$$

$$+\,^{60\,000}C_4 \left(\frac{39\,999}{40\,000}\right)^{59\,996} \left(\frac{1}{40\,000}\right)^4 \qquad\qquad \text{4 cases}$$

It is messy but you can evaluate it on your calculator. It comes out to be

$$0.223 + 0.335 + 0.251 + 0.126 + 0.047 = 0.981$$

(The figures are written to three decimal places but more places were used in the calculation.)

So the probability of 5 or more cases in a year is $1 - 0.981 = 0.019$. It is unlikely but certainly could happen, see figure 2.1.

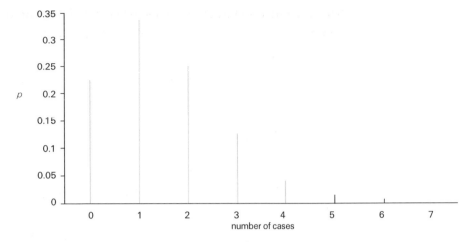

Figure 2.1 *Probability distribution* $B(60\,000, \frac{1}{40\,000})$

Note

Two other points are worth making. First, the binomial model assumes the trials are independent. If this disease is at all infectious, that certainly would not be the case. Second, there is no evidence at all to link this disease with *Avonford Chemicals*. There are many other possible explanations.

Approximating the binomial terms

Although it was possible to do the calculation using results derived from the binomial distribution, it was distinctly cumbersome. In this section you will see how the calculations can be simplified, a process which turns out to be unexpectedly profitable. The work that follows depends upon the facts that the event is rare but there are many opportunities for it to occur: that is, p is small and n is large.

Start by looking at the first term, the probability of 0 cases of the disease. This is

$$\left(\frac{39\,999}{40\,000}\right)^{60\,000} = k, \text{ a constant}$$

Now look at the next term, the probability of 1 case of the disease. This is

$$^{60\,000}C_1 \left(\frac{39\,999}{40\,000}\right)^{59\,999} \left(\frac{1}{40\,000}\right)$$

$$= \frac{60\,000 \times \left(\frac{39\,999}{40\,000}\right)^{60\,000} \times \left(\frac{40\,000}{39\,999}\right)}{40\,000}$$

$$= k \times \frac{60\,000}{39\,999} \approx k \times \frac{60\,000}{40\,000} = k \times 1.5$$

Now look at the next term, the probability of 2 cases of the disease. This is

$$^{60\,000}C_2 \times \left(\frac{39\,999}{40\,000}\right)^{59\,998} \times \left(\frac{1}{40\,000}\right)^2$$

$$= \frac{60\,000 \times 59\,999}{2 \times 1} \times \left(\frac{39\,999}{40\,000}\right)^{60\,000} \times \left(\frac{40\,000}{39\,999}\right)^2 \times \left(\frac{1}{40\,000}\right)^2$$

$$= \frac{k \times 60\,000 \times 59\,999}{2 \times 1 \times 39\,999 \times 39\,999} \approx \frac{k \times 60\,000 \times 60\,000}{2 \times 40\,000 \times 40\,000} = k \times \frac{(1.5)^2}{2}$$

Proceeding in this way leads to the following probability distribution for the number of cases of the disease:

Number of cases	0	1	2	3	4	$\cdots$
Probability	k	$k \times 1.5$	$\dfrac{k \times (1.5)^2}{2!}$	$\dfrac{k \times (1.5)^3}{3!}$	$\dfrac{k \times (1.5)^4}{4!}$	$\cdots$

Since the sum of the probabilities $= 1$,

$$k + k \times 1.5 + k \times \frac{(1.5)^2}{2!} + k \times \frac{(1.5)^3}{3!} + k \times \frac{(1.5)^4}{4!} + \cdots = 1$$

$$k\left[1 + 1.5 + \frac{(1.5)^2}{2!} + \frac{(1.5)^3}{3!} + \frac{(1.5)^4}{4!} + \cdots\right] = 1$$

The terms in the square brackets form a well known series in pure mathematics, the exponential series e^x.

$$e^x = 1 + x + \frac{x^2}{2!} + \frac{x^3}{3!} + \frac{x^4}{4!} + \cdots$$

Since $k \times e^{1.5} = 1$, $k = e^{-1.5}$.

This gives the probability distribution for the number of cases of the disease:

Number of cases	0	1	2	3	4	$\cdots$
Probability	$e^{-1.5}$	$e^{-1.5}1.5$	$e^{-1.5}\dfrac{(1.5)^2}{2!}$	$e^{-1.5}\dfrac{(1.5)^3}{3!}$	$e^{-1.5}\dfrac{(1.5)^4}{4!}$	$\cdots$

and in general for r cases the probability is $e^{-1.5}\dfrac{(1.5)^r}{r!}$

Accuracy

These expressions are clearly much simpler than those involving binomial coefficients. How accurate are they? The following table compares the results from the two methods, given to six decimal places.

No. of cases	Probability	
	Exact binomial method	Approximate method
0	0.223 126	0.223 130
1	0.334 697	0.334 695
2	0.251 025	0.251 021
3	0.125 512	0.125 511
4	0.047 066	0.047 067

You will see that the agreement is very good; there are no differences until the sixth decimal places.

The Poisson distribution

What started out as a search for an easy way to calculate terms in a binomial distribution has ended up with terms which are so different that they are seen as a completely different distribution, called the *Poisson distribution*.

In the example the distribution had mean 1.5 and the general term was given by $P(X = r) = e^{-1.5} \dfrac{(1.5)^r}{r!}$ where the discrete random variable X denotes the number of cases of the disease.

This can be generalised to the Poisson distribution with mean λ (pronounced 'lambda') for which $P(X = r) = e^{-\lambda} \dfrac{\lambda^r}{r!}$

Notes

1. The mean, λ, is usually called the population parameter.

2. e is the base of natural logarithms. e $= 2.718\,28\ldots$; like π, it does not terminate. e^x is a function on your calculator. It is sometimes written $\exp(x)$.

3. The upper case letter X is a random variable, and the lower case r is a particular value that X can take.

4. The shape of the Poisson distribution depends on the value of the parameter, λ. If λ is small the distribution has positive skew, but as λ increases the distribution becomes progressively more symmetrical, see figure 2.2.

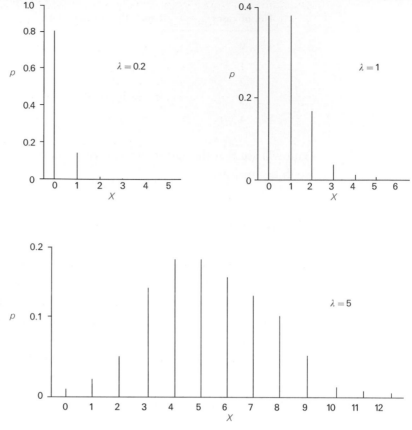

Figure 2.2 *The shape of the Poisson distribution for (a) $\lambda = 0.2$ (b) $\lambda = 1$ (c) $\lambda = 5$*

In the previous few pages, you have been introduced to the Poisson distribution as a very good approximation to the binomial distribution. However, it is much more significant than that. It is a distribution in its own right which can be used to model situations which are clearly not binomial, where you cannot state the number of trials or the probabilities of success and failure.

Conditions under which the Poisson distribution may be used

The Poisson distribution is generally thought of as the probability distribution for the number of occurrences of a *rare event*.

The conditions under which it may be used are as follows:

1. AS AN APPROXIMATION TO THE BINOMIAL DISTRIBUTION

The Poisson distribution may be used as an approximation to the binomial distribution, $B(n, p)$, when

(i) n is large
(ii) p is small (and so the event is rare).

In addition you would not usually want to use the Poisson distribution for large values of np, so in practice (but not in theory) a third condition may be applied:

(iii) np is not so large that the Poisson probabilities are more difficult to calculate than their binomial counterparts.

In the case of the disease, n was large at 60 000, p small at $\frac{1}{40\,000}$. In addition np, 1.5, was not large.

It is also necessary that the trials are random and independent; otherwise the distribution to be approximated would not be binomial in the first place.

2. AS A DISTRIBUTION IN ITS OWN RIGHT

There are many situations in which events happen singly and the mean number of occurrences is known (or can easily be found) but it is not possible, or even meaningful, to give values to the number of trials, n, or the probability of success, p. You can, for example, find the mean number of goals per team in league football matches, or the mean number of telephone calls received per minute at an exchange, but you cannot say how many goals are not scored or telephone calls not received. The concept of a trial, with success or failure as possible outcomes, is not appropriate to these situations. They may, however, be modelled by the Poisson distribution, provided that

(i)   the occurrences are random
(ii) the occurrences are independent
(iii) there is a known, constant overall mean rate for the occurrences.

It would be unusual to use the Poisson distribution with a population mean greater than about 20, because easier methods exist.

EXAMPLE 2.1

The number of defects in a wire cable can be modelled by the Poisson distribution with a mean of four defects per kilometre.

What is the probability that a single kilometre of wire will have exactly two defects?

SOLUTION

Let $X =$ the number of defects per kilometre

$$P(X = 2) = e^{-4}\frac{4^2}{2!}$$
$$= 0.147$$

Notation

The notation '$X \sim$ Poisson (6)' means that the random variable X has a Poisson distribution with parameter (or mean) 6.

So the probability that X takes the value 8 is given by $e^{-6} \dfrac{6^8}{8!}$ and the probability that it takes a general value r by $e^{-6} \dfrac{6^r}{r!}$

Calculating Poisson distribution probabilities

In the example about the disease afflicting Avonford, you had to work out $P(X \geqslant 5)$. To do this you used $P(X \geqslant 5) = 1 - P(X \leqslant 4)$ which saved you having to work out all the probabilities for 5 or more occurrences and adding them together. Such calculations can take a long time even though the terms eventually get smaller and smaller, so that after some time you will have gone far enough for the accuracy you require and may stop.

However, suppose you had $X \sim$ Poisson (8) and wanted to find $P(X \leqslant 7)$.

$$P(X \leqslant 7) = P(X = 0) + P(X = 1) + P(X = 2) + \cdots + P(X = 7)$$

This involves summing eight probabilities and so is itself rather tedious. Here are two ways of cutting down on the amount of work, and so on the time you take.

1. RECURRENCE RELATIONS

Recurrence relations allow you to use the term you have obtained to work out the next one. For the Poisson distribution with parameter λ,

$P(X = 0) = e^{-\lambda}$ You must use your calculator to find this term.

$P(X = 1) = e^{-\lambda}\lambda = \lambda P(X = 0)$ Multiply the previous term by λ.

$P(X = 2) = e^{-\lambda}\dfrac{\lambda^2}{2!} = \dfrac{\lambda}{2}P(X = 1)$

Multiply the previous term by $\dfrac{\lambda}{2}$.

$P(X = 3) = e^{-\lambda}\dfrac{\lambda^3}{3!} = \dfrac{\lambda}{3}P(X = 2)$

Multiply the previous term by $\dfrac{\lambda}{3}$.

$P(X = 4) = e^{-\lambda}\dfrac{\lambda^4}{4!} = \dfrac{\lambda}{4}P(X = 3)$

Multiply the previous term by $\dfrac{\lambda}{4}$.

In general, you can find $P(X = r)$ by multiplying your previous probability, $P(X = r - 1)$, by $\frac{\lambda}{r}$. You would expect to hold the latest value on your calculator and keep a running total in the memory.

Setting this out on paper with $\lambda = 1.5$ (the figure from the example about the disease, see page 35) gives these figures:

No. of cases, r	Conversion	$P(X = r)$	Running total, $P(X \leqslant r)$
0		0.223 130	0.223 130
	$\times 1.5$		
1		0.334 695	0.557 825
	$\times \dfrac{1.5}{2}$		
2		0.251 021	0.808 846
	$\times \dfrac{1.5}{3}$		
3		0.125 511	0.934 357
	$\times \dfrac{1.5}{4}$		
4		0.047 067	0.981 424

2. CUMULATIVE POISSON PROBABILITY TABLES

You can also find probabilities like $P(X \leqslant 4)$ by using *cumulative Poisson probability tables.* You can see how to do this by looking at the extract from the tables below. For $\lambda = 1.5$ and $x = 4$ this gives you the answer 0.9814.

x \ λ	1.00	1.10	1.20	1.30	1.40	1.50	1.60	1.70	1.80	1.90
0	0.3679	0.3329	0.3012	0.2725	0.2466	0.2231	0.2019	0.1827	0.1653	0.1496
1	0.7358	0.6990	0.6626	0.6268	0.5918	0.5578	0.5249	0.4932	0.4628	0.4337
2	0.9197	0.9004	0.8795	0.8571	0.8335	0.8088	0.7834	0.7572	0.7306	0.7037
3	0.9810	0.9743	0.9662	0.9569	0.9463	0.9344	0.9212	0.9068	0.8913	0.8747
4	0.9963	0.9946	0.9923	0.9893	0.9857	0.9814	0.9763	0.9704	0.9636	0.9559
5	0.9994	0.9990	0.9985	0.9978	0.9968	0.9955	0.9940	0.9920	0.9896	0.9868
6	0.9999	0.9999	0.9997	0.9996	0.9994	0.9991	0.9987	0.9981	0.9974	0.9966
7	1.0000	1.0000	1.0000	0.9999	0.9999	0.9998	0.9997	0.9996	0.9994	0.9992
8	.....	.....	.....	1.000	1.000	1.000	1.000	0.9999	0.9999	0.9998
9	.....	.....	.....	.....	.....	.....	.....	1.0000	1.0000	1.0000

Figure 2.3

If you wanted to find $P(X \geqslant 5)$, you would use

$$P(X \geqslant 5) = 1 - P(X \leqslant 4)$$
$$= 1 - 0.9814$$
$$= 0.0186$$

The variance of the Poisson distribution

If you think of the Poisson distribution as an approximation to the binomial, $B(n, p)$, when p is small and n is large, you can find an expression for the variance by approximating that for the binomial distribution.

For the binomial distribution, $B(n, p)$:
mean $= np$ and variance $= np(1 - p) = npq$.

When p is small, $q = 1 - p$ is approximately equal to 1.

Writing $q = 1$ in   variance $= npq$
gives   variance $= np$

You will see that this is just the same as the mean, giving the somewhat surprising result that for a Poisson distribution

variance $=$ mean

Both variance and mean are given the symbol λ and called the *population parameter*.

So if you have some data where the mean and variance are very similar then you may possibly be able to use the Poisson distribution as a model. In real life you will not find the mean and variance to be exactly the same. Remember that you are using the Poisson distribution as a model and it is unlikely to be a perfect fit.

Notes

1. You should realise that this is not a rigorous proof of the expression for the variance of the Poisson distribution. For that you would need to start, not with the binomial distribution, but with the Poisson probability:

$$P(X = r) = e^{-\lambda} \frac{\lambda^r}{r!}$$

This proof is somewhat longer and is given in the Appendix on page 135.

2. You may have wondered earlier why the term *parameter* was used rather than *mean* and why it was given the symbol λ rather than μ. Since the variance and mean are both the same, either of them can be used to tell you all you want to know about a Poisson distribution and so either is a parameter. To say that either one of the mean, μ, or variance, σ^2, is the parameter would be to give one of them greater status than the other. So a new symbol, λ, is used.

EXAMPLE 2.2

It is known that nationally one person in a thousand is allergic to a particular chemical used in making a wood preservative. A firm that makes this wood preservative employs 500 people in one of its factories.

(i) What is the probability that more than two people at the factory are allergic to the chemical?

(ii) What assumption are you making?

SOLUTION

(i) Let X be the number of people in a random sample of 500 who are allergic to the chemical.

$$X \sim B(500, 0.001) \qquad n = 500 \quad p = 0.001$$

Since n is large and p is small, the Poisson approximation to the binomial is appropriate.

$$\lambda = np$$
$$= 500 \times 0.001$$
$$= 0.5$$

Consequently
$$P(X = r) = e^{-\lambda} \frac{\lambda^r}{r!}$$
$$= e^{-0.5} \frac{0.5^r}{r!}$$

$$P(X > 2) = 1 - P(X \leqslant 2)$$
$$= 1 - [P(X = 0) + P(X = 1) + P(X = 2)]$$
$$= 1 - \left[e^{-0.5} + e^{-0.5}0.5 + e^{-0.5} \frac{0.5^2}{2} \right]$$
$$= 1 - [0.6065 + 0.3033 + 0.0758]$$
$$= 1 - 0.9856$$
$$= 0.014$$

This figure could have been found in cumulative Poisson probability tables.

(ii) The assumption made is that people with the allergy are just as likely to work in the factory as those without the allergy. In practice this seems rather unlikely: you would not stay in a job that made you unhealthy

EXAMPLE 2.3

Jasmit is considering buying a telephone answering machine. He has one for five days' free trial and finds that 22 messages are left on it. Assuming that this is typical of the use it will get if he buys it, find:

(i) the mean number of messages per day
(ii) the probability that on one particular day there will be exactly six messages
(iii) the probability that on one particular day there will be more than six messages.

SOLUTION

(i) Converting the total for five days to the mean for a single day gives

$$\text{daily mean} = \tfrac{22}{5} = 4.4 \text{ messages per day}$$

(ii) Calling X the number of messages per day,

$$P(X = 6) = e^{-4.4}\,\frac{4.4^6}{6!}$$
$$= 0.124$$

(iii) Using the cumulative Poisson probability tables gives

$$P(X \leqslant 6) = 0.8436$$

and so

$$P(X > 6) = 1 - 0.8436$$
$$= 0.1564$$

EXERCISE 2A

1 If $X \sim$ Poisson (2), calculate **(i)** $P(X=1)$ **(ii)** $P(X=4)$.

2 If $Z \sim$ Poisson (2.5), calculate **(i)** $P(Z=0)$ **(ii)** $P(Z=3)$ **(iii)** $P(Z=5)$.

3 If $W \sim$ Poisson (3), calculate
   (i) $P(W=0)$ **(ii)** $P(W=1)$ **(iii)** $P(W=2)$
   (iv) $P(W \leqslant 2)$ **(v)** $P(W>2)$.

4 If $X \sim$ Poisson (4.4), calculate **(i)** $P(X \leqslant 3)$ **(ii)** $P(X>3)$.

5 The number of wombats which are killed on a particular stretch of road in
   Australia in any one day can be modelled by a P(0.4) random variable.
   Calculate the probability that exactly two wombats are killed on a given day
   on this stretch of road.

   [Cambridge]

6 The number of cars passing a house in a residential road between 10 am and
   11 am on a weekday is a random variable, X. Give a condition under which X
   may be modelled by a Poisson distribution.

   Suppose that $X \sim$ P(3.4). Calculate $P(X \geqslant 4)$.

   [Cambridge]

7 The number of night calls to a fire station in a small town can be modelled
   by a Poisson distribution with mean 4.2 per night. Find the probability that
   on a particular night there will be three or more calls to the fire station.

   State what needs to be assumed about the calls to the fire station in order to
   justify a Poisson model.

   [Cambridge]

8 A typesetter makes 1500 mistakes in a book of 500 pages. On how many pages would you expect to find **(i)** 0 **(ii)** 1 **(iii)** 2 **(iv)** 3 or more mistakes? State any assumptions in your workings.

9 During the 1998–99 season Homerton F.C. scored the following numbers of goals in their league matches:

Goals	0	1	2	3	4	5	6
Matches	7	11	10	8	3	1	2

(i) How many league matches did Homerton play that season?

(ii) Calculate the mean and variance of the number of goals per match.

(iii) Using the mean you calculated in part (i) as the parameter in the Poisson distribution, calculate the probabilities of $0, 1, \ldots, 5, 6$ or more goals in any match.

(iv) In how many matches would you expect Homerton to have scored $0, 1, 2, \ldots, 6$ or more goals, according to the Poisson model?

(v) State, with reasons, whether you consider this to be a good model.

10 In a country the mean number of deaths per year from lightning strike is 2.2.

(i) Find the probabilities of 0, 1, 2 and more than 2 deaths from lightning strike in any particular year.

In a neighbouring country, it is found that one year in twenty nobody dies from lightning strike.

(ii) What is the mean number of deaths per year in that country from lightning strike?

11 A machine that puts an anchovy stuffing into olives fails to get the stuffing into 1 in every 50 olives. Find the probability that in a tub of 100 stuffed olives

(i) all the olives are correctly stuffed

(ii) there is exactly 1 unstuffed olive.

The olives are sold in packs of 250 tubs.

(iii) How many tubs in such a pack would you expect to have 4 or more unstuffed olives?

(iv) One person in a thousand complains if there are 4 or more unstuffed olives in a tub. How many complaints a year would you expect from a shop that orders 6 packs a month?

12 A manufacturer of rifle ammunition tests a large consignment for accuracy by firing 500 batches, each of 20 rounds, from a fixed rifle at a target. Those rounds that fall outside a marked circle on the target are classified as *misses*. For each batch of 20 rounds the number of misses is counted.

Misses, X	0	1	2	3	4	5	6–20
Frequency	230	189	65	15	0	1	0

(i) Estimate the mean number of misses per batch.

(ii) Use your mean to estimate the probability of a batch producing 0, 1, 2, 3, 4 and 5 misses using the Poisson distribution as a model.

(iii) Use your answers to part (ii) to estimate expected frequencies of 0, 1, 2, 3, 4 and 5 misses per batch in 500 batches and compare your answers with those actually found.

(iv) Do you think the Poisson distribution is a good model for this situation?

13 A survey in a town's primary schools has indicated that 5% of the pupils have severe difficulties with reading. If the primary school pupils were allocated to the secondary schools at random, estimate the probability that a secondary school with an intake of 200 pupils will receive

(i) no more than 8 pupils with severe reading difficulties

(ii) more than 20 pupils with severe reading difficulties.

14 Fanfold paper for computer printers is made by putting perforations every 30 cm in a continuous roll of paper. A box of fanfold paper contains 2000 sheets. State the length of the continuous rolls from which the box of paper is produced.

The manufacturers claim that faults occur at random and at an average rate of 1 per 240 metres of paper. State an appropriate distribution for the number of faults per box of paper. Find the probability that a box of paper has no faults and also the probability that it has more than 4 faults.

Two copies of a report which runs to 100 sheets per copy are printed on this sort of paper. Find the probability that there are no faults in either copy of the report and also the probability that just one copy is faulty.

[MEI]

15 A firm investigated the number of employees suffering injuries whilst at work. The results recorded below were obtained for a 52-week period:

Number of employees injured in a week	0	1	2	3	4 or more
Number of weeks	31	17	3	1	0

Give reasons why one might expect this distribution to approximate to a Poisson distribution. Evaluate the mean and variance of the data and explain why this gives further evidence in favour of a Poisson distribution.

Using the calculated value of the mean, find the theoretical frequencies of a Poisson distribution for the number of weeks in which 0, 1, 2, 3, 4 or more employees were injured.

16 350 raisins are put into a mixture which is well stirred and made into 100 small buns. Estimate how many of these buns will
  (i) be without raisins
  (ii) contain five or more raisins.

In a second batch of 100 buns, exactly one has no raisins in it.
  (iii) Estimate the total number of raisins in the second mixture.

17 A count was made of the number of red blood corpuscles in each of the 64 compartments of a haemocytometer with the following results:

Number of corpuscles	2	3	4	5	6	7	8	9	10	11	12	13	14
Frequency		1	5	4	9	10	10	8	6	4	3	2	1	1

Estimate the mean and variance of the number of red blood corpuscles per compartment. Explain how the values you have obtained support the view that those data are a sample from a Poisson population.

Write down an expression for the theoretical frequency with which compartments containing five red blood corpuscles should be found, assuming this to be obtained from a Poisson population with mean 7. Evaluate this frequency to two decimal places.

[MEI]

18 At a busy intersection of roads, accidents requiring the summoning of an ambulance occur with a frequency, on average, of 1.8 per week. These accidents occur randomly, so that it may be assumed that they follow a Poisson distribution.
  (i) Calculate the probability that there will not be an accident in a given week.
  (ii) Calculate the smallest integer n such that the probability of more than n accidents in a week is less than 0.02.
  (iii) Calculate the probability that there will not be an accident in a given fortnight.
  (iv) Calculate the largest integer k such that the probability that there will not be an accident in k successive weeks is greater than 0.0001.

[AEB]

19 A ferry takes cars and small vans on a short journey from an island to the mainland. On a representative sample of weekday mornings, the numbers of vehicles, X, on the 8 am sailing were as follows:

20	24	24	22	23		21	20	22	23	22
21	21	22	21	23		22	20	22	20	24

  (i) Show that X does not have a Poisson distribution.

In fact 20 of the vehicles belong to commuters who use that sailing of the ferry every weekday morning. The random variable Y is the number of vehicles other than those 20 who are using the ferry.

(ii) Investigate whether Y may reasonably be modelled by a Poisson distribution.

The ferry can take 25 vehicles on any journey.

(iii) On what proportion of days would you expect at least one vehicle to be unable to travel on this particular sailing of the ferry because there was no room left and so have to wait for the next one?

20 A traffic survey is being undertaken on a main road to determine whether or not a pedestrian crossing should be installed. On five successive days, from Monday to Friday, the hour between 8 am and 9 am was split up into 30-second intervals and the number of vehicles passing a certain point in each of these intervals was recorded.

The random variable X represents the number of cars travelling *from* the town centre per 30-second interval. For the 600 observations the mean and variance were 3.1 and 3.27 respectively.

(i) Explain why X might be modelled by a Poisson distribution.

(ii) Using the sample mean as an estimate for the Poisson parameter, calculate the probability of recording exactly 3 vehicles travelling *from* the town centre in a 30-second interval.

(iii) Calculate the probability of recording at least 6 vehicles travelling from the town centre in a 60-second interval.

[MEI]

INVESTIGATIONS

The Poisson distribution provides a remarkably accurate model for many situations. Here are some investigations for you to try, but before you start there are a number of points to keep in mind.

1. MODELLING

Remember that if your results do not give a Poisson distribution, then that tells you it is not a good model. It does not mean that your experiment has failed, merely that the model is inappropriate.

2. INDEPENDENCE

Each occurrence must be independent. If, for example, you collect data on the numbers of cars passing a point just beyond some traffic lights, say every 30 seconds, the data will not be independent.

3. DATA COLLECTION

Data collection for a statistics investigation may be less active than for other investigations. You can, for instance, collect data for a statistics investigation sitting at your desk with a newspaper in front of you.

Collect data on one of the following:

(i)   The number of goals scored by teams in the Football League on a single Saturday.

(ii) The number of telephone calls coming into a switchboard in a one-minute period. (You may wish to vary the time period to, say, five minutes if the calls are infrequent.)

(iii) The number of cars passing a suitable point in a one-minute period.

(iv) Data relating to a situation of your own choice which you think could be modelled well by the Poisson distribution.

When you have collected the data, go through the following steps.

1. Work out the mean and the variance and check that they are approximately equal.
2. Use the mean to work out the Poisson probability distribution and a suitable set of expected frequencies.
3. Compare these expected frequencies with your observations.

Note

In the first three cases, you know the number of events that have occurred, but not how many have not occurred, so you cannot use the binomial distribution. It makes no sense to ask 'How many goals were not scored?' You can estimate the mean, np, but not n or p on their own.

The sum of two or more Poisson distributions

It is often the case that you wish to add two or more Poisson distributions together.

EXAMPLE 2.4

A rare disease causes the death, on average, of 3.8 people per year in England, 0.8 in Scotland and 0.5 in Wales. As far as is known the disease strikes at random and cases are independent of one another.

What is the probability of 7 or more deaths from the disease on the British mainland (i.e. England, Scotland and Wales together) in any year?

SOLUTION

Notice first that:

(a) P(7 or more deaths) $= 1 -$ P(6 or fewer deaths)

(b) each of the three distributions fulfils the conditions for it to be modelled by the Poisson distribution.

You could find the probability of 6 or fewer deaths by listing all the different ways they could be allocated between the three countries (e.g. England 3, Scotland 1, Wales 2) and working out the probabilities of all of them, using the three individual Poisson distributions. That would be very time-consuming indeed.

It is much easier to add the three distributions together and treat the result as a single Poisson distribution.

The overall mean is given by $\quad$ 3.8 $\quad + \quad$ 0.8 $\quad + \quad$ 0.5 $\quad = \quad$ 5.1

$\qquad\qquad\qquad\qquad\qquad\qquad$ England $\qquad$ Scotland $\qquad$ Wales $\qquad$ Total

giving an overall distribution of Poisson (5.1).

The probability of 6 or fewer deaths is then found from cumulative Poisson probability tables to be 0.7474.

So the probability of 7 or more deaths is given by $1 - 0.7474 = 0.2526$

Notes

1. You may only add Poisson distributions in this way if they are independent of each other.

2. The proof of the validity of adding Poisson distributions in this way is given in the Appendix on page 137.

EXAMPLE 2.5

On a lonely Highland road in Scotland, cars are observed passing at the rate of 6 per day, and lorries at the rate of 2 per day. On the road is an old cattle grid which will soon need repair. The local works department decide that if the probability of more than 15 vehicles per day passing is less than 1% then the repairs to the cattle grid can wait until next spring, otherwise it will have to be repaired before the winter.

When will the cattle grid have to be repaired?

SOLUTION

Let C be the number of cars per day, L be the number of lorries per day, and V be the number of vehicles per day.

$$V = L + C$$

Assuming that a car or a lorry passing along the road is a random event and the two are independent:

$$C \sim \text{Poisson } (6), \quad L \sim \text{Poisson } (2)$$
$$\text{and so} \quad V \sim \text{Poisson } (6 + 2)$$
$$= V \sim \text{Poisson } (8)$$

From cumulative Poisson probability tables $P(V \leqslant 15) = 0.9918$.

The required probability is $\quad P(V > 15) = 1 - P(V \leqslant 15)$
$$= 1 - 0.9918$$
$$= 0.0082$$

This is just less than 1% and so the repairs are left until spring.

The modelling of this situation raises a number of questions.

1 Is it true that a car or lorry passing along the road is a random event, or are some of these regular users, like the lorry collecting the milk from the farms along the road? If, say, three of the cars and one lorry are regular daily users, what effect does this have on the calculation?

2 Is it true that every car or lorry travels independently of every other one?

3 There are no figures for bicycles or motorcycles or other vehicles. Why might this be so?

EXERCISE 2B

1 Betty drives along a 50-mile stretch of motorway 5 days a week 50 weeks a year. She takes no notice of the 70 mph speed limit and, when the traffic allows, travels between 95 and 105 mph. From time to time she is caught by the police and fined but she estimates the probability of this happening on any day is $\frac{1}{300}$. If she gets caught three times within three years she will be disqualified from driving. Use Betty's estimates of probability to answer the following questions.

(i) What is the probability of her being caught exactly once in any year?

(ii) What is the probability of her being caught less than three times in three years?

(iii) What is the probability of her being caught exactly three times in three years?

Betty is in fact caught one day and decides to be somewhat cautious, reducing her normal speed to between 85 and 95 mph. She believes this will reduce the probability of her being caught to $\frac{1}{500}$.

(iv) What is the probability that she is caught less than twice in the next three years?

2 Motorists in a particular part of the Highlands of Scotland have a choice between a direct route and a one-way scenic detour. It is known that on average one in forty of the cars on the road will take the scenic detour. The road engineer wishes to do some repairs on the scenic detour. He chooses a time when he expects 100 cars an hour to pass along the road.

Find the probability that, in any one hour,
(i) no cars (ii) at most 4 cars
will turn on to the scenic detour.

(iii) Between 10.30 am and 11.00 am it will be necessary to block the road completely. What is the probability that no car will be delayed?

3 A sociologist claims that only 3% of all suitably qualified students from inner city schools go on to university. Use his claim and the Poisson approximation to the bionomial distribution to estimate the probability that in a randomly chosen group of 200 such students

(i) exactly five go to university

(ii) more than five go to university.

(iii) If there is at most a 5% chance that more than n of the 200 students go to university, find the lowest possible value of n.

Another group of 100 students is also chosen. Find the probability that

(iv) exactly five of each group go to university

(v) exactly ten of all the chosen students go to university.

[MEI]

4 At the hot drinks counter in a cafeteria both tea and coffee are sold. The number of cups of coffee sold per minute may be assumed to be a Poisson variable with mean 1.5 and the number of cups of tea sold per minute may be assumed to be an independent Poisson variable with mean 0.5.

(i) Calculate the probability that in a given one-minute period exactly one cup of tea and one cup of coffee are sold.

(ii) Calculate the probability that in a given three-minute period fewer than five drinks altogether are sold.

(iii) In a given one-minute period exactly three drinks are sold. Calculate the probability that these are all cups of coffee.

[Cambridge]

5 The numbers of lorry drivers and car drivers visiting an all-night transport cafe between 2 am and 3 am on a Sunday morning have independent Poisson distributions with means 5.1 and 3.6 respectively. Find the probabilities that, between 2 am and 3 am on any Sunday,

(i) exactly five lorry drivers visit the cafe

(ii) at least one car driver visits the cafe

(iii) exactly five lorry drivers and exactly two car drivers visit the cafe.

By using the distribution of the *total* number of drivers visiting the cafe, find the probability that exactly seven drivers visit the cafe between 2 am and 3 am on any Sunday. Given that exactly seven drivers visit the cafe between 2 am and 3 am on one Sunday, find the probability that exactly five of them are driving lorries.

[MEI]

6 A garage uses a particular spare part at an average rate of five per week. Assuming that usage of this spare part follows a Poisson distribution, find the probability that

(i) exactly five are used in a particular week

(ii) at least five are used in a particular week

(iii) exactly ten are used in a two-week period

(iv) at least ten are used in a two-week period

(v) exactly five are used in each of two successive weeks.

If stocks are replenished weekly, determine the number of spare parts which should be in stock at the beginning of each week to ensure that on average the stock will be insufficient on no more than one week in a 52-week year.

7 Small hard particles are found in the molten glass from which glass bottles are made. On average, 15 particles are found per 100 kg of molten glass. If a bottle contains one or more such particles it has to be discarded.

Suppose bottles of mass 1 kg are made. It is required to estimate the percentage of bottles that have to be discarded. Criticise the following 'answer': *Since the material for 100 bottles contains 15 particles, approximately 15% will have to be discarded.*

Making suitable assumptions, which should be stated, develop a correct argument using a Poisson model, and find the percentage of faulty 1 kg bottles to three significant figures.

Show that about 3.7% of bottles of mass 0.25 kg are faulty.

[MEI]

8 Weak spots occur at random in the manufacture of a certain cable at an average rate of 1 per 100 metres. If X represents the number of weak spots in 100 metres of cable, write down the distribution of X.

Lengths of this cable are wound on to drums. Each drum carries 50 metres of cable. Find the probability that a drum will have three or more weak spots.

A contractor buys five such drums. Find the probability that two have just one weak spot each and the other three have none.

[AEB]

9 A crockery manufacturer tests dinner plates by taking a large sample from each day's production. When all the machinery is set correctly, there are on average two faulty plates per sample but this number rises if any part of the process is incorrectly set.

When the machine is working correctly, what is the probability that a test will result in
(i) no faulty plates?
(ii) at most four faulty plates?

The company reset the machinery (a process which is expensive in time) if there are five or more faulty plates in a sample.
(iii) What is the probability that the machinery is reset unnecessarily?

The company decide to change the basis on which they make the decision to reset their machinery so that they will now do so if the total number of faulty plates in three consecutive samples is at least f.
(iv) Find the smallest value of f which gives a probability of less than 1% that the machinery will be reset unnecessarily.

(v) Given your value of f, find the probability that a set of three consecutive samples fails to indicate that the machinery needs resetting when the mean number of faults per sample has in fact risen to 3.0.

10 A petrol station has service areas on both sides of a motorway, one to serve east-bound traffic and the other for west-bound traffic. The number of east-bound vehicles arriving at the station in one minute has a Poisson distribution with mean 0.9, and the number of west-bound vehicles arriving in one minute has a Poisson distribution with mean 1.6, the two distributions being independent.

(i) Find the probability that in a one-minute period
  (a) no vehicles arrive
  (b) more than two vehicles arrive at this petrol station,
  giving your answers correct to three places of decimals.

Given that in a particular one-minute period three vehicles arrive, find
(ii) the probability that they are all from the same direction
(iii) the most likely combination of east-bound and west-bound vehicles.

[Cambridge]

11 The following table gives the number f_r of each of 519 equal time intervals in which r radioactive atoms decayed.

Number of decays, r	0	1	2	3	4	5	6	7	8	$\geqslant 9$
Observed number of intervals, f_r	11	41	73	105	107	82	55	28	9	8

Estimate the mean and variance of r.

Suggest, with justification, a theoretical distribution from which the data could be a random sample. Hence calculate expected values of f_r and comment briefly on the agreement between these and the observed values.

In the experiment each time interval was of length 7.5 s. In a further experiment, 1000 time intervals each of length 5 s are to be examined. Estimate the number of these intervals within which no atoms will decay.

12 A car hire firm has three cars, which it hires out on a daily basis. The number of cars demanded per day follows a Poisson distribution with mean 2.1.
(i) Find the probability that exactly two cars are hired out on any one day.
(ii) Find the probability that all cars are in use on any one day.
(iii) Find the probability that all cars are in use on exactly 3 days of a 5-day week.
(iv) Find the probability that exactly 10 cars are demanded in a 5-day week. Explain whether or not such a demand could always be met.
(v) It costs the firm £20 a day to run each car, whether it is hired out or not. The daily hire charge per car is £50. Find the expected daily profit.

[MEI]

13 In Abbotson town centre, the number of incidents of criminal damage reported to the police averages two per week.

(i) Explain why the Poisson distribution might be thought to be a suitable model for the number of incidents of criminal damage reported per week.

(ii) Find the probabilities of the following events, according to the Poisson model.

(a) Exactly two incidents are reported in a week.

(b) Two consecutive weeks are incident-free.

(c) More than ten incidents are reported in a period of four weeks.

[MEI]

14 The probability that I dial a wrong number when making a telephone call is 0.015. In a typical week I will make 50 telephone calls. Using a Poisson approximation to a binomial model find, correct to two decimal places, the probability that in such a week

(i) I dial no wrong numbers

(ii) I dial more than two wrong numbers.

Comment on the suitability of the binomial model and of the Poisson approximation.

[Cambridge]

15 A Christmas draw aims to sell 5000 tickets, 50 of which will each win a prize.

(i) A syndicate buys 200 tickets. Let X represent the number of tickets that win a prize.

(a) Justify the use of the Poisson approximation for the distribution of X.

(b) Calculate $P(X \leq 3)$.

(ii) Calculate how many tickets should be bought in order for there to be a 90% probability of winning at least one prize.

[Cambridge]

Historical note

Siméon Poisson was born in Pithiviers in France in 1781. Under family pressure he began to study medicine but after some time gave it up for his real interest, mathematics. For the rest of his life Poisson lived and worked as a mathematician in Paris. His contribution to the subject spanned a broad range of topics in both pure and applied mathematics, including integration, electricity and magnetism and planetary orbits as well as statistics. He was the author of between 300 and 400 publications and originally derived the Poisson distribution as an approximation to the binomial distribution.

When he was a small boy, Poisson had his hands tied by his nanny who then hung him from a hook on the wall so that he could not get into trouble while she went out. In later life he devoted a lot of time to studying the motion of a pendulum and claimed that this interest derived from his childhood experience swinging against the wall.

The Poisson distribution

2

1 THE POISSON PROBABILITY DISTRIBUTION

- If $X \sim$ Poisson (λ) the parameter $\lambda > 0$.

$$P(X = r) = e^{-\lambda} \frac{\lambda^r}{r!} \quad r \geq 0, \; r \text{ is an integer}$$
$$E(X) = \lambda$$
$$Var(X) = \lambda$$

2 CONDITIONS UNDER WHICH THE POISSON DISTRIBUTION MAY BE USED

- The Poisson distribution is generally thought of as the probability distribution for the number of occurrences of a *rare event*.

- *As a distribution in its own right*
  Situations in which the mean number of occurrences is known (or can easily be found) but it is not possible, or even meaningful, to give values to n or p may be modelled using the Poisson distribution provided that the occurrences are

—random
—independent.

- *As an approximation to the binomial distribution*
  The Poisson distribution may be used as an approximation to the binomial distribution, $B(n, p)$, when

— n is large
— p is small (and so the event is rare)
— np is not large.

It would be unusual to use the Poisson distribution with parameter, λ, greater than about 20.

3 THE SUM OF TWO POISSON DISTRIBUTIONS

If $X \sim$ Poisson (λ), $Y \sim$ Poisson (μ) and X and Y are independent

$$X + Y \sim \text{Poisson } (\lambda + \mu).$$

3 The normal distribution

The
normal
law of error
stands out in the
experience of mankind
as one of the broadest
generalisations of natural
philosophy. It serves as the
guiding instrument in researches
in the physical and social sciences
and in medicine, agriculture and engineering.
It is an indispensable tool for the analysis and the
interpretation of the basic data obtained by observation and experiment.

W. J. Youden

THE AVONFORD STAR

VILLAGERS GET GIANT BOBBY

The good people of Middle Fishbrook have special reason to be good these days. Since last week, their daily lives are being watched over by their new village bobby, Wilf 'Shorty' Harris.

At 195 cm, Wilf is the tallest policeman in the country. 'I don't expect any trouble', says Wilf. 'But I wouldn't advise anyone to tangle with me on a dark night.'

Seeing Wilf towering above me, I decided that most people would prefer not to put his words to the test.

Towering Bobby 'Shorty' Harris is bound to deter mischief in Middle Fishbrook

Wilf Harris is clearly exceptionally tall, but how much so? Is he one in a hundred, or a thousand or even a million? To answer that question you need to know the distribution of heights of adult British men.

The first point that needs to be made is that height is a continuous variable and not a discrete one. If you measure accurately enough it can take any value.

This means that it does not really make sense to ask 'What is the probability that somebody chosen at random has height exactly 195 cm?' The answer is zero.

However, you can ask questions like 'What is the probability that somebody chosen at random has height between 194 cm and 196 cm?' and 'What is the probability that somebody chosen at random has height at least 195 cm?'. When the variable is continuous, you are concerned with a range of values rather than a single value.

Like many other naturally occurring variables, the heights of adult men may be modelled by the normal distribution, shown below. You will see that this has a distinctive bell-shaped curve and is symmetrical about its middle. The curve is continuous as height is a continuous variable.

On figure 3.1, area represents probability so the shaded area to the right of 195 cm represents the probability that a randomly selected adult male is over 195 cm tall.

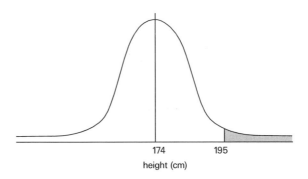

174   195

height (cm)

Figure 3.1

Before you can start to find this area, you must know the mean and standard deviation of the distribution, in this case about 174 cm and 7 cm respectively.

So Wilf's height is 195 cm − 174 cm = 21 cm above the mean, and that is

$$\frac{21}{7} = 3 \text{ standard deviations}$$

The number of standard deviations beyond the mean, in this case 3, is denoted by the letter z. Thus the shaded area gives the probability of obtaining a value of $z \geqslant 3$.

You find this area by looking up the value of $\Phi(z)$ when $z = 3$ in a normal distribution table of $\Phi(z)$ as shown in figure 3.2, and then calculating $1 - \Phi(z)$. (Φ is the Greek letter phi.)

$\Phi(3) = .9987$

z	.00	.01	.02	.03	.04	.05	.06	.07	.08	.09	ADD 1 2 3 4 5 6 7 8 9
0.0	.5000	5040	5080	5120	5160	5199	5239	5279	5319	5359	4 8 12 16 20 24 28 32 36
0.1	.5398	5438	5478	5517	5557	5596	5636	5675	5714	5753	4 8 12 16 20 24 28 32 35
0.2	.5793	5832	5871	5910	5948	5987	6026	6064	6103	6141	4 8 12 15 19 23 27 31 35
0.3	.6179	6217	6255	6293	6331	6368	6406	6443	6480	6517	4 8 11 15 19 23 26 30 34
0.4	.6554	6591	6628	6664	6700	6736	6772	6808	6844	6879	4 7 11 14 18 22 25 29 32
0.5	.6915	6950	6985	7019	7054	7088	7123	7157	7190	7224	3 7 10 14 17 21 24 27 31
0.6	.7257	7291	7324	7357	7389	7422	7454	7486	7517	7549	3 6 10 13 16 19 23 26 29
0.7	.7580	7611	7642	7673	7704	7734	7764	7794	7823	7852	3 6 9 12 15 18 21 24 27
0.8	.7881	7910	7939	7967	7995	8023	8051	8078	8106	8133	3 6 8 11 14 17 19 22 25
3.0	.9987	9987	9987	9988	9988	9989	9989	9989	9990	9990	
3.1	.9990	9991	9991	9991	9992	9992	9992	9992	9993	9993	*differences*
3.2	.9993	9993	9994	9994	9994	9994	9994	9995	9995	9995	*untrustworthy*
3.3	.9995	9995	9996	9996	9996	9996	9996	9996	9996	9997	
3.4	.9997	9997	9997	9997	9997	9997	9997	9997	9997	9998	

Figure 3.2 *Extract from tables of* $\Phi(z)$

This gives $\Phi(3) = 0.9987$, and so $1 - \Phi(3) = 0.0013$.

The probability of a randomly selected adult male being 195 cm or over is 0.0013. Slightly more than one man in a thousand is at least as tall as Wilf.

Using normal distribution tables

The function $\Phi(z)$ gives the area under the normal distribution curve to the *left* of the value z, that is the shaded area in figure 3.3 (it is the cumulative distribution function). The total area under the curve is 1, and the area given by $\Phi(z)$ represents the probability of a value smaller than z.

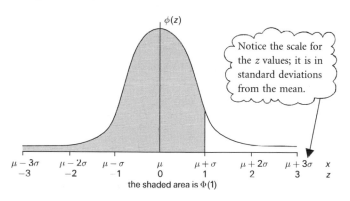

Notice the scale for the z values; it is in standard deviations from the mean.

the shaded area is $\Phi(1)$

Figure 3.3

If the variable X has mean μ and standard deviation σ then x, a particular value of X, is transformed into z by the equation

$$z = \frac{x - \mu}{\sigma}$$

z is a particular value of the variable Z which has mean 0 and standard deviation 1 and is the *standardised* form of the normal distribution.

	Actual distribution, X	Standardised distribution, Z
Mean	μ	0
Standard deviation	σ	1
Particular value	x	$z = \dfrac{x - \mu}{\sigma}$

Notice how lower case letters, x and z, are used to indicate particular values of the random variables, whereas upper case letters, X and Z, are used to describe or name those variables.

Normal distribution tables are easy to use but you should always make a point of drawing a diagram and shading the region you are interested in.

It is often helpful to know that in a normal distribution, roughly:

- 68% of the values lie within ± 1 standard deviation of the mean;

- 95% of the values lie within ± 2 standard deviations of the mean;

- 99.75% of the values lie within ± 3 standard deviations of the mean.

EXAMPLE 3.1

Assuming the distribution of the heights of adult men is normal, with mean 174 cm and standard deviation 7 cm, find the probability that a randomly selected adult man is

(i)   under 185 cm
(ii) over 185 cm
(iii) over 180 cm
(iv) between 180 cm and 185 cm
(v)   under 170 cm

giving answers to 2 significant figures.

SOLUTION

The mean height, $\mu = 174$

The standard deviation, $\sigma = 7$

(i)   The probability that an adult man selected at random is under 185 cm.

The area required is that shaded in the diagram.

$$x = 185 \, cm$$

and so $\quad z = \dfrac{185 - 174}{7} = 1.571$

$$\Phi(1.571) = 0.9419$$
$$= 0.94 \quad (2 \, sf)$$

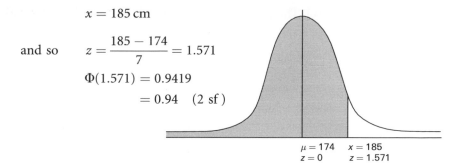

$\mu = 174 \quad x = 185$
$z = 0 \quad\quad z = 1.571$

Figure 3.4

Answer: The probability that an adult man selected at random is under 185 cm is 0.94.

(ii) The probability that an adult man selected at random is over 185 cm.

The area required is the complement of that for part (i).

$$\text{Probability} = 1 - \Phi(1.571)$$
$$= 1 - 0.9419$$
$$= 0.0581$$
$$= 0.058 \quad (2 \, sf)$$

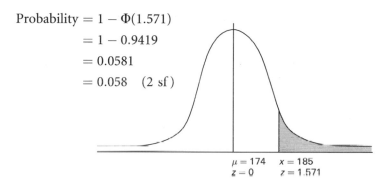

$\mu = 174 \quad x = 185$
$z = 0 \quad\quad z = 1.571$

Figure 3.5

Answer: The probability that an adult man selected at random is over 185 cm is 0.058.

(iii) The probability that an adult man selected at random is over 180 cm.

$x = 180 \quad$ and so $\quad z = \dfrac{180 - 174}{7} = 0.857$

The area required $= 1 - \Phi(0.857)$
$$= 1 - 0.8042$$
$$= 0.1958$$
$$= 0.20 \quad (2 \, sf)$$

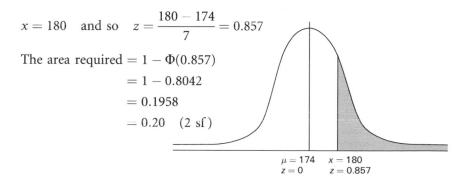

$\mu = 174 \quad x = 180$
$z = 0 \quad\quad z = 0.857$

Figure 3.6

Answer: The probability that an adult man selected at random is over 180 cm is 0.20.

(iv) The probability that an adult man selected at random is between 180 cm and 185 cm.

The required area is shown in the diagram. It is

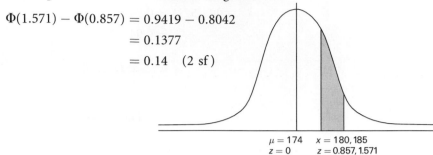

$$\Phi(1.571) - \Phi(0.857) = 0.9419 - 0.8042$$
$$= 0.1377$$
$$= 0.14 \quad (2 \text{ sf})$$

$\mu = 174 \qquad x = 180, 185$
$z = 0 \qquad\quad z = 0.857, 1.571$

Figure 3.7

Answer: The probability that an adult man selected at random is over 180 cm but under 185 cm is 0.14.

(v) The probability that an adult man selected at random is under 170 cm.

In this case $x = 170$

and so $\qquad z = \dfrac{170 - 174}{7} = -0.571$

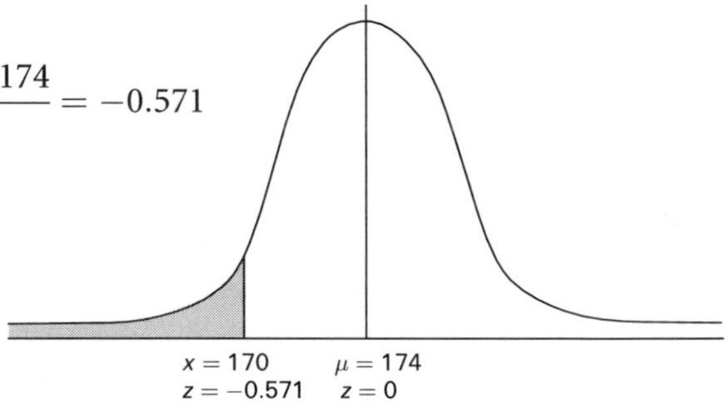

$x = 170 \qquad \mu = 174$
$z = -0.571 \qquad z = 0$

Figure 3.8

However when you come to look up $\Phi(-0.571)$, you will find that only positive values of z are given in your tables. You overcome this problem by using the symmetry of the normal curve. The area you want in this case is that to the left of -0.571 and this is clearly just the same as that to the right of $+0.571$.

So $\qquad \Phi(-0.571) = 1 - \Phi(0.571)$
$$= 1 - 0.716 = 0.284$$
$$= 0.28 \quad (2 \text{ sf})$$

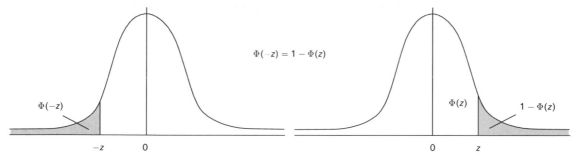

Figure 3.9

Answer: The probability that an adult man selected at random is under 170 cm is 0.28.

The normal curve

All normal curves have the same basic shape, so that by scaling the two axes suitably you can always fit one normal curve exactly on top of another one.

The curve for the normal distribution with mean μ and standard deviation σ (i.e. variance σ^2) is given by the function $\phi(x)$ in

$$\phi(x) = \frac{1}{\sigma\sqrt{2\pi}}\, e^{-\frac{1}{2}\left(\frac{x-\mu}{\sigma}\right)^2}$$

The notation $N(\mu, \sigma^2)$ is used to describe this distribution. The mean, μ, and standard deviation, σ (or variance, σ^2), are the two parameters used to define the distribution. Once you know their values, you know everything there is to know about the distribution. The standardised variable Z has mean 0 and variance 1, so its distribution is $N(0, 1)$.

After the variable X has been transformed to Z using $z = \frac{x-\mu}{\sigma}$ the form of the curve (now standardised) becomes

$$\psi(z) = \frac{1}{\sqrt{2\pi}}\, e^{-\frac{1}{2}z^2}$$

However, the exact shape of the normal curve is often less useful than the area underneath it, which represents a probability. For example, the probability that $Z \leqslant 2$ is given by the shaded area in figure 3.10.

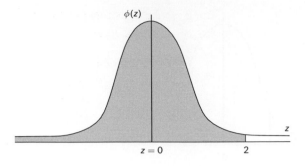

Figure 3.10

Easy though it looks, the function $\phi(z)$ cannot be integrated algebraically to find the area under the curve, only by a numerical method. The values found by doing so are given as a table and this area function is called $\Phi(z)$.

EXAMPLE 3.2

Skilled operators make a particular component for an engine. The company believes that the time taken to make this component may be modelled by the normal distribution with mean 95 minutes and standard deviation 4 minutes.

Assuming the company's belief to be true find the probability that the time taken to make one of these components, selected at random, was

(i)   over 97 minutes
(ii) under 90 minutes
(iii) between 90 and 97 minutes.

Sheila believes that the company is allowing too long for the job and invites them to time her. They find that only 10% of the components take her over 90 minutes to make, and that 20% take her less than 70 minutes.

(iv) Estimate the mean and standard deviation of the time Sheila takes.

SOLUTION

According to the company $\mu = 95$ and $\sigma = 4$ so the distribution is $N(95, 4^2)$.

(i)   The probability that a component required over 97 minutes

$$z = \frac{97 - 95}{4} = 0.5$$

The probability is represented by the shaded area and given by

$$1 - \Phi(0.5) = 1 - 0.6915$$
$$= 0.3085$$
$$= 0.31 \quad (2 \text{ dp})$$

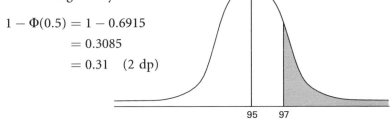

Figure 3.11

Answer: The probability it took the operator over 97 minutes to manufacture a randomly selected component is 0.31.

(ii) The probability that a component required under 90 minutes

$$z = \frac{90 - 95}{4} = -1.25$$

The probability is represented by the shaded area and given by

$$1 - \Phi(1.25) = 1 - 0.8944$$
$$= 0.1056$$
$$= 0.11 \quad (2 \text{ dp})$$

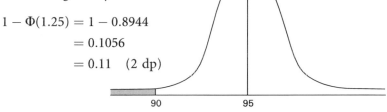

Figure 3.12

Answer: The probability it took the operator under 90 minutes to manufacture a randomly selected component is 0.11.

(iii) The probability that a component required between 90 and 97 minutes.

The probability is represented by the shaded area and given by

$$1 - 0.1056 - 0.3085 = 0.5859$$
$$= 0.59 \quad (2 \text{ dp})$$

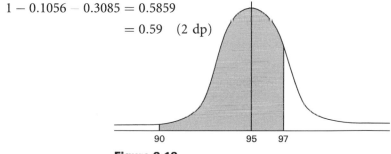

Figure 3.13

Answer: The probability it took the operator between 90 and 97 minutes to manufacture a randomly selected component is 0.59.

(iv) Estimate the mean and standard deviation of the time Sheila takes.

The question has now been put the other way round. You have to infer the mean, μ, and standard deviation, σ, from the areas under different parts of the graph.

10% take her 90 minutes or more. This means that the shaded area is 0.1.

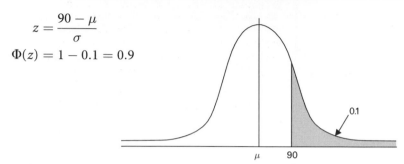

$$z = \frac{90 - \mu}{\sigma}$$

$$\Phi(z) = 1 - 0.1 = 0.9$$

Figure 3.14

You must now use the table of the inverse normal function, $\Phi^{-1}(p) = z$ which tells you $z = 1.282$.

The inverse normal function – values of $\Phi^{-1}(p) = z$

p	.00	.01	.02	.03	.04	.05	.06	.07	.08	.09
.50	.0000	.0025	.0050	.0075	.0100	.0125	.0150	.0175	.0201	.0226
.51	.0251	.0276	.0301	.0326	.0351	.0376	.0401	.0426	.0451	.0476
.52	.0502	.0527	.0552	.0577	.0602	.0627	.0652	.0677	.0702	.0728
.89	1.227	1.232	1.237	1.243	1.248	1.254	1.259	1.265	1.270	1.276
.90	1.282	1.287	1.293	1.299	1.305	1.311	1.317	1.323	1.329	1.335
.91	1.341	1.347	1.353	1.360	1.366	1.372	1.379	1.385	1.392	1.398
.92	1.405	1.412	1.419	1.426	1.433	1.440	1.447	1.454	1.461	1.468

$\Phi^{-1}(0.9) = 1.282$

Figure 3.15 *Extract from the inverse normal distribution table*

If you do not have the table of the inverse normal function, you can find the same answer by using the table of $\Phi(z) = p$ in reverse.

z	.00	.01	.02	.03	.04	.05	.06	.07	.08	.09	1	2	3	4	5	6	7	8	9
0.0	.5000	5040	5080	5120	5160	5199	5239	5279	5319	5359	4	8	12	16	20	24	28	32	36
0.1	.5398	5438	5478	5517	5557	5596	5636	5675	5714	5753	4	8	12	16	20	24	28	32	35
0.2	.5793	5832	5871	5910	5948	5987	6026	6064	6103	6141	4	8	12	15	19	23	27	31	35
0.3	.6179	6217	6255	6293	6331	6368	6406	6443	6480	6517	4	8	11	15	19	23	26	30	34
0.4	.6554	6591	6628	6664	6700	6736	6772	6808	6844	6879	4	7	11	14	18	22	25	29	32
0.5	.6915	6950	6985	7019	7054	7088	7123	7157	7190	7224	3	7	10	14	17	21	24	27	31
0.6	.7257	7291	7324	7357	7389	7422	7454	7486	7517	7549	3	6	10	13	16	19	23	26	29
0.7	.7580	7611	7642	7673	7704	7734	7764	7794	7823	7852	3	6	9	12	15	18	21	24	27
0.8	.7881	7910	7939	7967	7995	8023	8051	8078	8106	8133	3	6	8	11	14	17	19	22	25
0.9	.8159	8186	8212	8238	8264	8289	8315	8340	8365	8389	3	5	8	10	13	15	18	20	23
1.0	.8413	8438	8461	8485	8508	8531	8554	8577	8599	8621	2	5	7	9	12	14	16	18	21
1.1	.8643	8665	8686	8708	8729	8749	8770	8790	8810	8830	2	4	6	8	10	12	14	16	19
1.2	.8849	8869	8888	8907	8925	8944	8962	8980	8997	9015	2	4	6	7	9	11	13	15	16
1.3	.9032	9049	9066	9082	9099	9115	9131	9147	9162	9177	2	3	5	6	8	10	11	13	14
1.4	.9192	9207	9222	9236	9251	9265	9279	9292	9306	9319	1	3	4	6	7	8	10	11	13

$\Phi^{-1}(0.9) = 1.2815$

Figure 3.16 *Extract from tables of Φ(z)*

Returning to the problem you now know that

$$\frac{90 - \mu}{\sigma} = 1.282 \text{ simplifying to } 90 - \mu = 1.282\sigma.$$

The second piece of information, that 20% of components took Sheila under 70 minutes, is illustrated in this diagram.

$$z = \frac{70 - \mu}{\sigma}$$

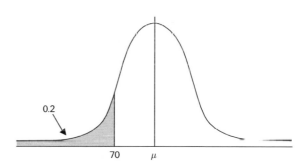

Figure 3.17

(z has a negative value in this case, the point being to the left of the mean.)

$$\Phi(z) = 0.2$$

and so, by symmetry,

$$\Phi(-z) = 1 - 0.2 = 0.8$$

Using the table of the inverse normal function gives

$$-z = 0.8416 \quad \text{or} \quad z = -0.8416$$

This gives a second equation for μ and σ.

$$\frac{70 - \mu}{\sigma} = -0.8416 \text{ simplifying to } 70 - \mu = -0.8416\sigma$$

You now solve the two simultaneous equations

$$90 - \mu = 1.282\sigma$$
$$70 - \mu = -0.8416\sigma$$

Subtract $\quad 20 \quad = 2.1236\sigma \qquad \sigma = 9.418 = 9.4 \quad \text{(1 dp)}$

$$\text{and} \qquad \mu = 77.926 = 77.9 \quad \text{(1 dp)}$$

Answer: Sheila's mean time is 77.9 minutes with standard deviation 9.4 minutes.

EXERCISE 3A

1   The distribution of the heights of 18-year-old girls may be modelled by the normal distribution with mean 162.5 cm and standard deviation 6 cm. Find the probability that the height of a randomly selected 18-year-old girl is
   (i)   under 168.5 cm
   (ii) over 174.5 cm
   (iii) between 168.5 and 174.5 cm.

2   A pet shop has a tank of goldfish for sale. All the fish in the tank were hatched at the same time and their weights may be taken to be normally distributed with mean 100 g and standard deviation 10 g. Melanie is buying a goldfish and is invited to catch the one she wants in a small net. In fact the fish are much too quick for her to be able to catch any particular one and the fish which she eventually nets is selected at random. Find the probability that its weight is
   (i)   over 115 g
   (ii) under 105 g
   (iii) between 105 and 115 g.

3   When he makes instant coffee, Tony puts a spoonful of powder into a mug. The weight of coffee in grams on the spoon may be modelled by the normal distribution with mean 5 and standard deviation 1. If he uses more than 6.5 g Julia complains that it is too strong and if he uses less than 4 g she tells him it is too weak. Find the probability that he makes the coffee
   (i)   too strong
   (ii) too weak
   (iii) all right.

4   When a butcher takes an order for a Christmas turkey, he asks the customer what weight in kilograms the bird should be. He then sends his order to a turkey farmer who supplies birds of about the requested weight. For any particular weight of bird ordered, the error in kilograms may be taken to be normally distributed with mean 0 and standard deviation 0.75.

Mrs Jones orders a 10 kg turkey from the butcher. Find the probability that the one she gets is

(i)   over 12 kg

(ii) under 10 kg

(iii) within 0.5 kg of the weight she actually ordered.

5 A biologist finds a nesting colony of a previously unknown sea bird on a remote island. She is able to take measurements on 100 of the eggs before replacing them in their nests. She records their weights, w g, in this frequency table.

Weight, w	$25 < w \leqslant 27$	$27 < w \leqslant 29$	$29 < w \leqslant 31$	$31 < w \leqslant 33$	$33 < w \leqslant 35$	$35 < w \leqslant 37$
Frequency	2	13	35	33	17	0

(i)   Find the mean and standard deviation of these data.

(ii) Assuming the weights of the eggs for this type of bird are normally distributed and that their mean and standard deviation are the same as those of this sample, find how many eggs you would expect to be in each of these categories.

(iii) Do you think the assumption that the weights of the eggs are normally distributed is reasonable?

6 A normally distributed random variable X has mean 20.0 and variance 4.15. Find the probability that $18.0 < X < 21.0$.

[Cambridge]

7 The length of life of a certain make of tyre is normally distributed about a mean of 24 000 km with a standard deviation of 2500 km.

(i)   What percentage of such tyres will need replacing before they have travelled 20 000 km?

(ii) As a result of improvements in manufacture, the length of life is still normally distributed, but the proportion of tyres failing before 20 000 km is reduced to 1.5%.

  (a) If the standard deviation has remained unchanged, calculate the new mean length of life.

  (b) If, instead, the mean length of life has remained unchanged, calculate the new standard deviation.

[MEI]

8 A machine is set to produce nails of lengths 10 cm, with standard deviation 0.05 cm. The lengths of the nails are normally distributed.

(i)   Find the percentage of nails produced between 9.95 cm and 10.08 cm in length.

The machine's setting is moved by a careless apprentice with the consequence that 16% of the nails are under 5.2 cm in length and 20% are over 5.3 cm.

(ii) Find the new mean and standard deviation.

9 The weights of eggs, measured in grams, can be modelled by a N(85.0, 36.0) distribution. Eggs are classified as large, medium or small, where a large egg weighs 90.0 grams or more, and 25% of eggs are classified as small. Calculate

(i) the percentage of eggs which are classified as large

(ii) the maximum weight of a small egg.

[Cambridge]

10 The concentration by volume of methane at a point on the centre line of a jet of natural gas mixing with air is distributed approximately normally with mean 20% and standard deviation 7%. Find the probabilities that the concentration

(i) exceeds 30%

(ii) is between 5% and 15%.

(iii) In another similar jet, the mean concentration is 18% and the standard deviation is 5%. Find the probability that in at least one of the jets the concentration is between 5% and 15%.

[MEI]

11 In a particular experiment, the length of a metal bar is measured many times. The measured values are distributed approximately normally with mean 1.340 m and standard deviation 0.021 m. Find the probabilities that any one measured value

(i) exceeds 1.370 m.

(ii) lies between 1.310 m and 1.370 m

(iii) lies between 1.330 m and 1.390 m.

(iv) Find the length l for which the probability that any one measured value is less than l is 0.1.

[MEI]

12 Each weekday a man goes to work by bus. His arrival time at the bus stop is normally distributed with standard deviation 3 minutes. His mean arrival time is 8.30 am. Buses leave promptly every 5 minutes at 8.21 am, 8.26 am, etc. Find the probabilities that he catches the buses at

(i) 8.26 am   **(ii)** 8.31 am   **(iii)** 8.36 am

assuming that he always gets on the first bus to arrive.

(iv) The man is late for work if he catches a bus after 8.31 am. What mean arrival time would ensure that, on average, he is not late for work more than one day in five? [Assume that he cannot change the standard deviation of his arrival time and give your answer to the nearest 10 s.]

[MEI]

13 A machine is used to fill cans of soup with a nominal volume of 0.450 litres. Suppose that the machine delivers a quantity of soup which is normally distributed with mean μ litres and standard deviation σ litres. Given that $\mu = 0.457$ and $\sigma = 0.004$, find the probability that a randomly chosen can contains less than the nominal volume.

It is required by law that no more than 1% of cans contain less than the nominal volume. Find

(i) the least value of μ which will comply with the law when $\sigma = 0.004$

(ii) the greatest value of σ which will comply with the law when $\mu = 0.457$.

[MEI]

14 A factory is lit by a large number of electric light bulbs whose lifetimes are modelled by a normal distribution with mean 1000 hours and standard deviation 110 hours. Operating conditions require that all bulbs are on continuously.

(i) What proportions of bulbs have lifetimes that

  (a) exceed 950 hours **(b)** exceed 1050 hours?

(ii) Given that a bulb has already lasted 950 hours, what is the probability that it will last a further 100 hours?

Give all answers correct to 3 decimal places.

The factory management is to adopt a policy whereby all bulbs will be replaced periodically after a fixed interval.

(iii) To the nearest day, how long should this interval be if, on average, 1% of the bulbs are to burn out between successive replacement times?

[MEI]

15 A factory produces a very large number of steel bars. The lengths of these bars are normally distributed with 33% of them measuring 20.06 cm or more and 12% of them measuring 20.02 cm or less.

Write down two simultaneous equations for the mean and standard deviation of the distribution and solve to find values to 4 significant figures. Hence estimate the proportion of steel bars which measure 20.03 cm or more.

The bars are acceptable if they measure between 20.02 cm and 20.08 cm. What percentage are rejected as being outside the acceptable range?

[MEI]

16 Two firms, Goodline and Megadelay, produce delay lines for use in communications. The delay time for a delay line is measured in nanoseconds (ns).

(i) The delay times for the output of Goodline may be modelled by a normal distribution with mean 283 ns and standard deviation 8 ns. What is the probability that the delay time of one line selected at random from Goodline's output is between 275 and 286 ns?

(ii) It is found that, in the output of Megadelay, 10% of the delay times are less than 274.6 ns and 7.5% are more than 288.2 ns. Again assuming a normal distribution, calculate the mean and standard deviation of the delay times for Megadelay. Give your answers correct to 3 significant figures.

[Cambridge]

17 The diameters D of screws made in a factory are normally distributed with mean 1 mm. Given that 10% of the screws have diameters greater than 1.04 mm, find the standard deviation correct to 3 significant figures, and hence show that about 2.7% of the screws have diameters greater than 1.06 mm.

Find, correct to 2 significant figures,

(i) the number d for which 99% of the screws have diameters that exceed d mm

(ii) the number e for which 99% of the screws have diameters that do not differ from the mean by more than e mm.

[MEI]

18 A firm makes two different brands of similar electronic components, A and B. The life of brand A has mean 23 hours, standard deviation 2 hours; the life of brand B has mean 25 hours, standard deviation 5 hours. Their lives are assumed to be distributed according to a normal probability model.

(i) Which brand is more likely to break down over a period of

(a) 22 hours   **(b)** 20 hours?

Replacing such a component in the course of a certain job causes expensive delay, so a new component is fitted before starting the job. If the component lasts for x hours, it ensures a profit $£F(x)$, where

$$F(x) = \begin{cases} -100, & 0 < x < 20; \\ 20, & 20 \leqslant x < 25; \\ 40, & x \geqslant 25. \end{cases}$$

(ii) Which brand gives the greater expected profit?

[SMP]

19 A machine produces crankshafts whose diameters are normally distributed with mean 5 cm and standard deviation 0.03 cm. Find the percentage of crankshafts it will produce whose diameters lie between 4.95 cm and 4.97 cm.

What is the probability that two successive crankshafts will both have a diameter in this interval?

Crankshafts with diameters outside the interval 5 ± 0.05 cm are rejected. If the mean diameter of the machine's production remains unchanged, to what must the standard deviation be reduced if only 4% of the production is to be rejected?

[MEI]

20 A soft drinks dispenser delivers lemonade into a cup when a coin is inserted into the machine. The amount of lemonade delivered is normally distributed with mean 260 ml and standard deviation 10 ml. The nominal amount of lemonade in a cup is 250 ml. The capacity of the cup is 275 ml.

(i)   What is the probability that the cup overflows?

(ii) What is the probability that the amount of lemonade in the cup is at least 250 ml but does not overflow?

(iii) On one occasion my friends and I purchase five such cups of lemonade. What is the probability that not more than one cup contains less than 250 ml?

(iv) Some customers have complained that the proportion of cups giving short measure is too high. The standard deviation of the amount of lemonade delivered per cup is fixed, but the mean can be altered. What value would you recommend so that no more than 5% of cups contain less than 250 ml of lemonade?

[MEI]

Modelling discrete situations

Although the normal distribution applies strictly to a continuous variable, it is also common to use it in situations where the variable is discrete providing that:

- the distribution is approximately normal; this requires that the steps in its possible values are small compared with its standard deviation;
- *continuity corrections* are applied where appropriate.

The meaning of the term continuity correction is explained in the following example.

EXAMPLE 3.3

The result of an Intelligence Quotient (IQ) test is an integer score, X. Tests are designed so that X has a mean value of 100 with standard deviation 15. A large number of people have their IQs tested. What proportion of them would you expect to have IQs measuring between 106 and 110 (inclusive)?

SOLUTION

Although the random variable X is an integer and hence discrete, the steps of 1 in its possible values are small compared with the standard deviation of 15. So it is reasonable to treat it as if it is continuous.

If you assume that an IQ test is measuring innate, natural intelligence (rather than the results of learning), then it is reasonable to assume a normal distribution.

If you draw the probability distribution function for the discrete variable X it looks like figure 3.18. The area you require is the total of the five bars representing 106, 107, 108, 109 and 110.

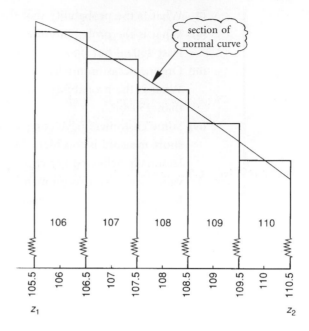

Figure 3.18

The equivalent section of the normal curve would run not from 106 to 110 but from 105.5 to 110.5, as you can see in the diagram. When you change from the discrete scale to the continuous scale, the numbers 106, 107 etc. no longer represent the whole intervals, just their centre points.

So the area you require under the normal curve is given by $\Phi(z_2) - \Phi(z_1)$
where $z_1 = \dfrac{105.5 - 100}{15}$ and $z_2 = \dfrac{110.5 - 100}{15}$.
This is $\Phi(0.7000) - \Phi(0.3667)$
$$= 0.7580 - 0.6431 = 0.1149$$

Answer: The proportion of IQs between 106 and 110 (inclusive) should be approximately 11%.

In this calculation, both end values needed to be adjusted to allow for the fact that a continuous distribution was being used to approximate a discrete one. These adjustments, $106 \to 105.5$ and $110 \to 110.5$, are called continuity corrections. Whenever a discrete distribution is approximated by a continuous one a continuity correction may need to be used.

You must always think carefully when applying a continuity correction. Should the corrections be added or subtracted? In this case 106 and 110 are inside the required area and so any value (like 105.7 or 110.4) which would round to them must be included. It is often helpful to draw a sketch to illustrate the region you want, like the one in figure 3.18.

If the region of interest is given in terms of inequalities, you should look carefully to see whether they are inclusive ($\leqslant$ or $\geqslant$) or exclusive ($<$ or $>$). For example $20 \leqslant X \leqslant 30$ becomes $19.5 \leqslant X \leqslant 30.5$ whereas $20 < X < 30$ becomes $20.5 \leqslant X \leqslant 29.5$.

Two particularly common situations are when the normal distribution is used to approximate the binomial and the Poisson distributions.

Approximating the binomial distribution

You may use the normal distribution as an approximation for the binomial, $B(n, p)$ (where n is the number of trials each having probability p of success) when

1 n is large
2 p is not too close to 0 or 1.

These conditions ensure that the distribution is reasonably symmetrical and not skewed away from either end, see figure 3.19.

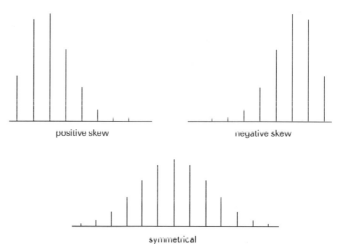

Figure 3.19

The parameters for the normal distribution are then

$$\text{Mean:} \qquad \mu = np$$
$$\text{Variance:} \quad \sigma^2 = npq$$

so that it can be denoted by $N(np, npq)$.

EXAMPLE 3.4

This is a true story. During voting at a by-election, an exit poll of 1700 voters indicated that 50% of people had voted for the Labour party candidate. When the votes were counted it was found that he had in fact received 57% support.

850 of the 1700 people interviewed said they had voted Labour but 57% of 1700 is 969, a much higher number. What went wrong? Is it possible to be so far out just by being unlucky and asking the wrong people?

SOLUTION

The situation of selecting a sample of 1700 people and asking them if they voted for one party or not is one that is modelled by the binomial distribution, in this case B(1700, 0.57).

In theory you could multiply out $(0.43 + 0.57t)^{1700}$ and use that to find the probabilities of getting $0, 1, 2, \ldots, 850$ Labour voters in your sample of 1700. In practice such a method would be impractical because of the work involved.

What you can do is to use a normal approximation. The required conditions are fulfilled: at 1700, n is certainly not small; $p = 0.57$ is near neither 0 nor 1.

The parameters for the normal approximation are given by

$$\text{Mean, } \mu = np = 1700 \times 0.57 = 969$$
$$\text{S.D., } \sigma = \sqrt{npq} = \sqrt{1700 \times 0.57 \times 0.43} = 20.4$$

You will see that the standard deviation, 20.4, is large compared with the steps of 1 in the possible values of Labour voters.

The probability of getting no more than 850 Labour voters, $P(X \leqslant 850)$, is given by $\Phi(z)$, where

$$z = \frac{850.5 - 969}{20.4} = -5.8$$

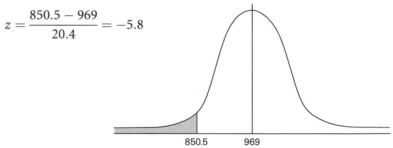

Figure 3.20

(Notice the continuity correction making 850 into 850.5.)

This is beyond the range of most tables and corresponds to a probability of about 0.000 01. The probability of a result as extreme as this is thus 0.000 02 (allowing for an equivalent result in the tail above the mean). It is clearly so unlikely that this was a result of random sampling that another explanation must be found.

❓ What do you think went wrong with the exit poll? Remember this really did happen.

Approximating the Poisson distribution

You may use the normal distribution as an approximation for the Poisson distribution, provided that its parameter (mean) λ is sufficiently large for the distribution to be reasonably symmetrical and not positively skewed.

As a working rule λ should be at least 10.

If $\lambda = 10$, mean $= 10$

and standard deviation $= \sqrt{10} = 3.16$.

A normal distribution is almost entirely contained within 3 standard deviations of its mean and in this case the value 0 is slightly more than 3 standard deviations away from the mean value of 10.

The parameters for the normal distribution are then

$$\text{Mean:} \quad \mu = \lambda$$
$$\text{Variance:} \quad \sigma^2 = \lambda$$

so that it can be denoted by $N(\lambda, \lambda)$.

(Remember that, for a Poisson distribution, mean $=$ variance.)

For values of λ larger than 10 the Poisson probability graph becomes less positively skewed and more bell-shaped in appearance thus making the normal approximation appropriate. Figure 3.21 shows the Poisson probability graph for the two cases $\lambda = 3$ and $\lambda = 25$. You will see that for $\lambda = 3$ the graph is positively skewed but for $\lambda = 25$ it is approximately bell-shaped.

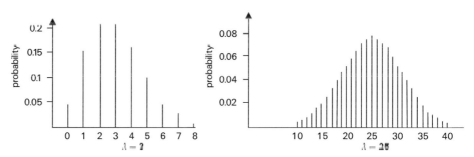

Figure 3.21

EXAMPLE 3.5

The annual number of deaths nationally from a rare disease, X, may be modelled by the Poisson distribution with mean 25. One year there are 31 deaths and it is suggested that the disease is on the increase.

What is the probability of 31 or more deaths in a year, assuming the mean has remained at 25?

SOLUTION

The Poisson distribution with mean 25 may be approximated by the normal distribution with parameters

Mean: 25

S.D.: $\sqrt{25} = 5$

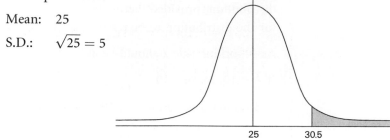

Figure 3.22

The probability of there being 31 or more deaths in a year, $P(X \geqslant 31)$, is given by $1 - \Phi(z)$, where

$$z = \frac{30.5 - 25}{5} = 1.1$$

(Note the continuity correction, replacing 31 by 30.5.)

The required area is $1 - \Phi(1.1)$
$$= 1 - 0.8643$$
$$= 0.1357$$

This is not a particularly low probability; it is quite likely that there would be that many deaths in any one year.

EXERCISE 3B

1 The intelligence of an individual is frequently described by a positive integer known as an IQ (intelligence quotient). The distribution of IQs amongst children of a certain age-group can be approximated by a normal probability model with mean 100 and standard deviation 15. Write a sentence stating what you understand about the age-group from the fact that $\Phi(2.5) = 0.994$.

A class of 30 children is selected at random from the age-group. Calculate (to 3 significant figures) the probability that at least one member of the class has an IQ of 138 or more.

[SMP]

2 A certain examination has a mean mark of 100 and a standard deviation of 15. The marks can be assumed to be normally distributed.
  (i) What is the least mark needed to be in the top 35% of pupils taking this examination?
  (ii) Between which two marks will the middle 90% of the pupils lie?
  (iii) 150 pupils take this examination. Calculate the number of pupils likely to score 110 or over.

[MEI]

3 25% of Flapper Fish have red spots, the rest blue spots. A fisherman nets 10 Flapper Fish. What are the probabilities that
  (i) exactly 8 have blue spots?
  (ii) at least 8 have blue spots?

A large number of samples, each of 100 Flapper Fish, are taken.
  (iii) What is
    (a) the mean?
    (b) the standard deviation of the number of red-spotted fish per sample?
  (iv) What is the probability of a sample of 100 Flapper Fish containing over 30 with red spots?

4 A fair coin is tossed 10 times. Evaluate the probability that exactly half of the tosses result in heads.

The same coin is tossed 100 times. Use the normal approximation to the binomial to estimate the probability that exactly half of the tosses result in heads. Also estimate the probability that more than 60 of the tosses result in heads.

Explain why a continuity correction is made when using the normal approximation to the binomial and the reason for the adoption of this correction.

[MEI]

5 State conditions under which a binomial probability model can be well approximated by a normal model.

X is a random variable with the distribution B$(12, 0.42)$.
  (i) Anne uses the binomial distribution to calculate the probability that $X < 4$ and gives 4 significant figures in her answer. What answer should she get?
  (ii) Ben uses a normal distribution to calculate an approximation for the probability that $X < 4$ and gives 4 significant figures in his answer. What answer should he get?
  (iii) Given that Ben's working is correct, calculate the percentage error in his answer.

[Cambridge]

6 During an advertising campaign, the manufacturers of Wolfitt (a dog food) claimed that 60% of dog owners preferred to buy Wolfitt.
  (i) Assuming that the manufacturer's claim is correct for the population of dog owners, calculate
    (a) using the binomial distribution
    (b) using a normal approximation to the binomial
    the probability that at least 6 of a random sample of 8 dog owners prefer to buy Wolfitt. Comment on the agreement, or disagreement, between your two values. Would the agreement be better or worse if the proportion had been 80% instead of 60%?

(ii) Continuing to assume that the manufacturer's figure of 60% is correct, use the normal approximation to the binomial to estimate the probability that, of a random sample of 100 dog owners, the number preferring Wolfitt is between 60 and 70 inclusive.

[MEI]

7 A multiple-choice examination consists of 20 questions, for each of which the candidate is required to tick as correct one of three possible answers. Exactly one answer to each question is correct. A correct answer gets 1 mark and a wrong answer gets 0 marks. Consider a candidate who has complete ignorance about every question and therefore ticks at random. What is the probability that he gets a particular answer correct? Calculate the mean and variance of the number of questions he answers correctly.

The examiners wish to ensure that no more than 1% of completely ignorant candidates pass the examination. Use the normal approximation to the binomial, working throughout to 3 decimal places, to establish the pass mark that meets this requirement.

[MEI]

8 A large box contains many plastic syringes, but previous experience indicates that 10% of the syringes in the box are defective. Five syringes are taken at random from the box. Use a binomial model to calculate, giving your answers correct to 3 decimal places, the probability that
(i) none of the five syringes is defective
(ii) at least two syringes out of the five are defective.

Discuss the validity of the binomial model in this context.

Instead of removing five syringes, 100 syringes are picked at random and removed. A normal distribution may be used to estimate the probability that at least 15 out of the 100 syringes are defective. Give a reason why it may be convenient to use a normal distribution to do this, and calculate the required estimate.

[Cambridge]

9 Recently our local Health Centre claimed that 95% of adults who have a flu jab during October do not catch flu during the next six months.

In parts (i) to (iv), assume that this claim is true and that people catch flu independently of each other.
(i) Find the probability that, of 25 adults given a flu jab, just 2 catch flu within the next six months.

During October, the Health Centre gave a flu jab to 184 adults.
(ii) Write down the parameter of an approximating Poisson distribution that describes the number of people who catch flu within six months. Explain briefly why this is a suitable approximation.

(iii) Find the probability of more than 10 of those given flu jabs catching flu.

(iv) Find the least integer k such that the probability of k or more adults catching flu is less than 5%.

(v) In fact, 20 of the adults given flu jabs in October catch flu during the next six months. What does this suggest about the assumptions made earlier?

[MEI]

10 A telephone exchange serves 2000 subscribers, and at any moment during the busiest period there is a probability of $\frac{1}{30}$ for each subscriber that he will require a line. Assuming that the needs of subscribers are independent, write down an expression for the probability that exactly N lines will be occupied at any moment during the busiest period.

Use the normal distribution to estimate the minimum number of lines that would ensure that the probability that a call cannot be made because all the lines are occupied is less than 0.01.

Investigate whether the total number of lines needed would be reduced if the subscribers were split into two groups of 1000, each with its own set of lines.

[MEI]

11 The number of cars per minute entering a multi-storey car park can be modelled by a Poisson distribution with mean 2. What is the probability that three cars enter during a period of one minute?

What are the mean and the standard deviation of the number of cars entering the car park during a period of 30 minutes? Use the normal approximation to the Poisson distribution to estimate the probability that at least 50 cars enter in any one 30-minute period.

[MEI]

12 State the mean and variance of the Poisson distribution. State under what circumstances the normal distribution can be used as an approximation to the Poisson distribution.

Readings, on a counter, of the number of particles emitted from a radioactive source in a time T seconds have a Poisson distribution with mean 250 T. A ten-second count is made. Find the probabilities of readings of **(i)** more than 2600 **(ii)** 2400 or more.

[JMB]

13 A drug manufacturer claims that a certain drug cures a blood disease on average 80% of the time. To check the claim, an independent tester uses the drug on a random sample of n patients. He decides to accept the claim if k or more patients are cured.

Assume that the manufacturer's claim is true.

(i) State the distribution of X, the number of patients cured. Find the probability that the claim will be accepted when 15 individuals are tested and k is set at 10.

A more extensive trial is now undertaken on a random sample of 100 patients.

(ii) State a suitable approximating distribution for X, and so estimate the probability that the claim will be rejected if k is set at 75.

(iii) Find the largest value of k such that the probability of the claim being rejected is no more than 1%.

[MEI]

14 A large computer system which is in constant operation requires an average of 30 service calls per year.

(i) State the average number of service calls per month, taking a month to be $\frac{1}{12}$ of a year. What assumptions need to be made for the Poisson distribution to be used to model the number of calls in a given month?

(ii) Use the Poisson distribution to find the probability that at least one service call is required in January. Obtain the probability that there is at least one service call in each month of the year.

(iii) The service contract offers a discount if the number of service calls in the year is 24 or fewer. Use a suitable approximating distribution to find the probability of obtaining the discount in any particular year.

[MEI]

The central limit theorem

The normal distribution occupies a central position in statistics. In this chapter you have met it in situations where it occurs naturally and also as an approximation to other distributions. Its use, however, goes well beyond this.

Statistical work often involves drawing conclusions from samples. In order to do so you need to understand the distributions of sample statistics, and in particular the sample mean. This is described by the central limit theorem:

> For samples of size n drawn from a distribution with mean μ and finite variance σ^2, the distribution of the sample mean is approximately $N\left(\mu, \dfrac{\sigma^2}{n}\right)$ for sufficiently large n.

Even if the underlying distribution is not normal, the distribution of the means of samples of a particular size drawn from it is approximately normal. The larger the sample size, the closer is this distribution to the normal.

So, even if the normal distribution never occurred in nature you would still need to know about it when drawing conclusions from sample data.

If the underlying distribution is normal, then the distribution of the sample mean is normal whatever the size of n.

The distribution of the scores on a die is uniform, with probability $\frac{1}{6}$ for each of the outcomes 1, 2, 3, 4, 5 and 6. It is not a normal distribution.

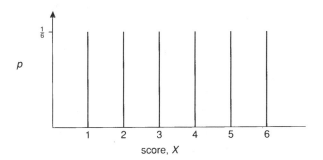

Figure 3.23

Throw five dice (or one die five times) and record the total score. This will be somewhere between 5 (5 ones) and 30 (5 sixes) and probably somewhere in the middle of the range. If it were, say, 16, then the sample mean would be $\frac{16}{5} = 3.2$. Mark this on a graph. Repeat the experiment many times, and watch the normal curve grow on your graph.

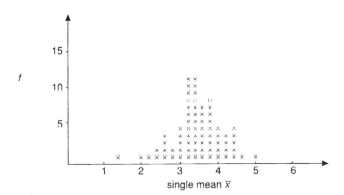

Figure 3.24

This experiment can obviously be conducted with different numbers of dice. The resulting mean will actually be a discrete variable (with steps of 0.2 in this case); the larger the number of dice the more accurate will be the ultimate curve, but the longer the experiment will take you. It can also be set up as a computer simulation.

Normal probability graph paper

There will be times when you will want to judge whether data you have collected could have come from a normal distribution. There are formal tests for establishing this but they are outside the scope of this book. A simple procedure is to use *normal probability graph paper*, as in the following example.

EXAMPLE 3.6   The weights of 50 adult female pet cats were measured as follows.

Weight (kg)	$2.0 < w \leqslant 3.0$	$3.0 < w \leqslant 4.0$	$4.0 < w \leqslant 5.0$	$5.0 < w \leqslant 6.0$	$6.0 < w \leqslant 7.0$
Frequency	2	9	15	12	6

Weight (kg)	$7.0 < w \leqslant 8.0$	$8.0 < w \leqslant 9.0$	$9.0 < w \leqslant 10.0$
Frequency	3	1	2

Do these data appear to be a sample from a normal distribution?

SOLUTION

The first step is to construct a cumulative frequency table and then to write the frequencies as percentage probabilities.

Weight	Frequency	Percentage
$\leqslant 2$	0	0
$\leqslant 3$	2	4
$\leqslant 4$	11	22
$\leqslant 5$	26	52
$\leqslant 6$	38	76
$\leqslant 7$	44	88
$\leqslant 8$	47	94
$\leqslant 9$	48	96
$\leqslant 10$	50	100

The percentage figures are then plotted on normal probability graph paper, as shown in figure 3.25, but without the first and last point (0% and 100%). If a straight line results, they would indeed seem to be a sample from a normal distribution.

You can see that these data do not give a straight line. It seems that the distribution of the weights of female pet cats is not normal. In fact if you look at the graphs in figure 3.26 you will see at a glance that there is considerable positive skew, presumably due to a number of very heavy cats, possibly fat creatures, spending their lives curled up by the fire!

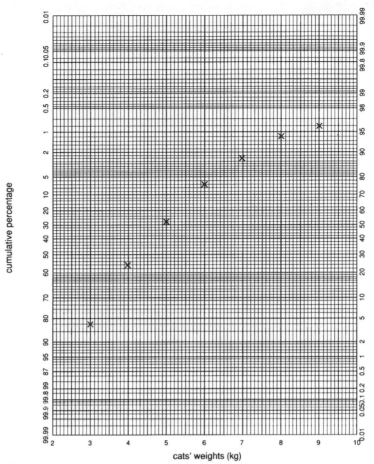

Figure 3.25

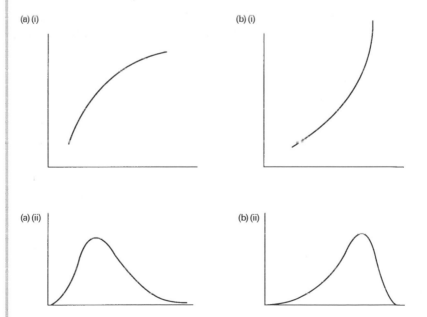

Figure 3.26 *(a) Positive skew (b) Negative skew.*
Drawn (i) on normal probability graph paper (ii) as a distribution

1 The normal distribution with mean μ and standard deviation σ is denoted by $N(\mu, \sigma^2)$.

2 This may be given in standardised form by using the transformation

$$z = \frac{x - \mu}{\sigma}$$

3 In the standardised form, $N(0, 1)$, the mean is 0, the standard deviation and variance both 1.

4 The standardised normal curve is given by

$$\Phi(z) = \frac{1}{\sqrt{2\pi}\,e^{-\frac{1}{2}z^2}}$$

5 The area to the left of the value z in the figure below, representing the probability of a value less than z, is denoted by $\Phi(z)$ and is read from tables.

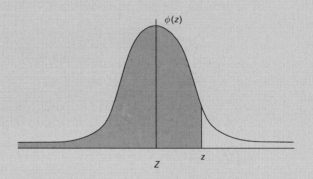

6 The normal distribution may be used to approximate suitable discrete distributions but continuity corrections are then required.

7 The binomial distribution $B(n, p)$ may be approximated by $N(np, npq)$, provided n is large and p is not close to 0 or 1.

8 The Poisson distribution Poisson (λ) may be approximated by $N(\lambda, \lambda)$, provided λ is about 10 or more.

4 Bivariate data

It is now proved beyond doubt that smoking is one of the leading causes of statistics.

John Peers

Ferguson to sign two new strikers

Tom Ferguson, the manager of Avonford Rovers Football Club, is set to sign two strikers in his bid to win promotion to the Assembly League next season. A buoyant Tom Ferguson told me this morning 'You have to score a lot of goals in this game and then the points will look after themselves. It's the same at all levels of the game, from the Premiership down.'

Tony Shields, from Walkden, has just recovered from a bunion operation and is the man Tom has in mind to lead the attack. At 195 cm tall he is formidable at set pieces and in the attacking half of the field. His strike partner will be Harry Gregory from Middle Fishbrook. Tom wants to sign Gregory before one of the big clubs come in and make him an offer. Ferguson told me, 'Gregory is only 160 cm tall but is like greased lightning in and around the box. People will probably call our new partnership Little and Large, but let's hope they score plenty of goals and then we'll be all right.'

Tony Shields clashes with Lumumba Athletic keeper Fred Weaver

---

? Do you agree with Tom Ferguson that if your team can score a lot of goals then the points total will look after itself?

Tom says it's the same principle from the Premiership down and so let us put the Premiership to the test. Here are the final positions and details of all clubs at the end of the 1998–99 season:

	P	W	D	L	F	A	Pts
Manchester United	38	22	13	3	80	37	79
Arsenal	38	22	12	4	59	17	78
Chelsea	38	20	15	3	57	30	75
Leeds	38	18	13	7	62	34	67
West Ham	38	16	9	13	46	53	57
Aston Villa	38	15	10	13	51	46	55
Liverpool	38	15	9	14	68	49	54
Derby	38	13	13	12	40	45	52
Middlesborough	38	12	15	11	48	54	51
Leicester	38	12	13	13	40	46	49
Tottenham	38	11	14	13	47	50	47
Sheffield Wednesday	38	13	7	18	41	42	46
Newcastle	38	11	13	14	48	54	46
Everton	38	11	10	17	42	47	43
Coventry	38	11	9	18	39	51	42
Wimbledon	38	10	12	16	40	63	42
Southampton	38	11	8	19	37	64	41
Charlton	38	8	12	18	41	56	36
Blackburn	38	7	14	17	38	52	35
Nottingham Forest	38	7	9	22	35	69	30

The graph in figure 4.1 illustrates the goals scored and points total for all 20 competing teams.

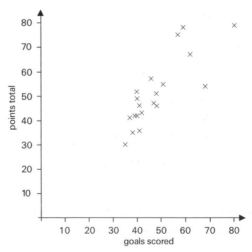

Figure 4.1 *Scatter diagram showing goals scored and points total for all teams in the Premiership 1998–99 season*

Looking at the spread of data points it does seem that the teams scoring many goals are the teams with the highest points totals.

The data in this example are a set of pairs of values for two variables, the goals scored and the points totals of all the teams in the 1998–99 football Premiership. This is an example of *bivariate data*, where each item in the population requires the values of two variables. The graph in figure 4.1 is called a *scatter diagram* and this is a common way of showing bivariate data.

If each point lies on a straight line, then there is said to be perfect *linear correlation* between the two variables. It is much more likely, however, that your data fall close to a straight line but not exactly on it. The better the fit, the higher the level of linear correlation.

The term *line of best fit* is used to describe a straight line drawn through a set of data points so as to fit them as closely as possible. There are several ways of determining such a line, according to what precisely is meant by a close fit.

Describing variables

DEPENDENT AND INDEPENDENT VARIABLES

The scatter diagram in figure 4.1 was drawn with the goals scored on the horizontal axis and the points total on the vertical axis. It was done that way to emphasise that the number of points is dependent upon the number of goals scored. (A team gains points as a result of scoring goals. It does not score goals as a result of gaining points.) It is normal practice to plot the *dependent* variable on the vertical axis and the *independent variable* on the horizontal axis.

Here are some more examples of dependent and independent variables.

Independent variable	Dependent variable
Number of people in a lift	Total weight of passengers
The amount of rain falling on a field whilst a crop is growing	The weight of the crop yielded
The number of people visiting a bar in an evening	The volume of beer sold

RANDOM AND NON-RANDOM VARIABLES

In the examples above both the variables have unpredictable values and so are *random*.

The same is true for the example about goals scored and points totals in football. Both variables are random variables, free to assume any of a particular set of discrete values in a given range.

Sometimes one or both of the variables is *controlled* so that the variable only assumes a set of predetermined values. Controlled variables are non-random. Here is an example:

Age of a rat in months	6	12	18	24	30
Memory quotient (on a particular scale)	16	22	25	25	23

In this example the age of the rat is a controlled variable. The tests were conducted every six months and not at random times so the age of the rat is a *non-random* variable.

Later in this chapter you will be asked to consider the nature of the variables you are dealing with in more detail.

Interpreting scatter diagrams

You can often judge if correlation is present just by looking at a scatter diagram.

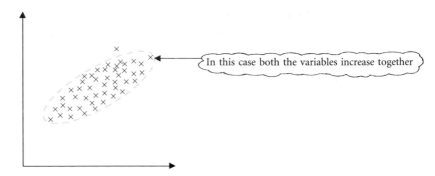

Figure 4.2 *Positive correlation*

Notice that in figure 4.2 almost all the observation points can be contained within an ellipse. This shape often arises when both variables are random. You should look for it before going on to do a calculation of Pearson's product moment correlation coefficient (see page 99). The narrower the elliptical profile, the greater the correlation.

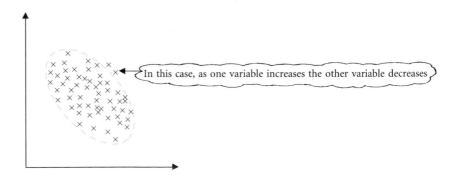

Figure 4.3 *Negative correlation*

In figure 4.3 the points again fall into an elliptical profile and this time there is negative correlation. The fatter ellipse in this diagram indicates weaker correlation than in the case shown in figure 4.2.

behaves

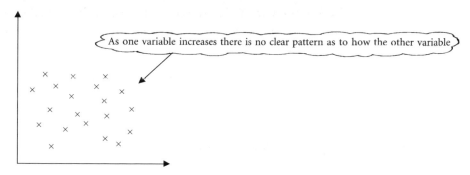

Figure 4.4 *No correlation*

In the case illustrated in figure 4.4 the points fall randomly in the (x, y) plane and there appears to be no association between the variables.

⚠ You should be aware of some distributions which at first sight appear to indicate linear correlation but in fact do not.

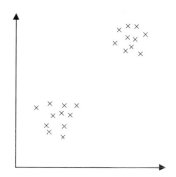

Figure 4.5 *Two islands*

This scatter diagram is probably showing two quite different groups, neither of them having any correlation.

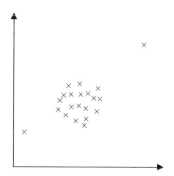

Figure 4.6 *Outliers*

This is a small data set with no correlation. However, the two outliers give the impression that there is positive linear correlation.

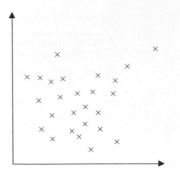

Figure 4.7 *A funnel-shaped distribution*

The bulk of these data have no correlation but a few items give the impression that there is correlation.

> *Note*
>
> In none of these three false cases is the distribution even approximately elliptical.

INVESTIGATIONS

There are many situations you can investigate for yourself. Here are two which you can carry out with a group of your friends.

DICE

Toss a pair of ordinary but distinguishable dice, A and B, 100 times and record the following data for each toss:

(a) the score on die A
(b) the total score on the two dice
(c) the difference between the scores on the two dice.

Plot three scatter diagrams to compare (a) and (b); (a) and (c); (b) and (c). Are you able to draw any conclusions about the relationships between the data?

BRAINS

Select a group of students of about the same age.

For each student, find the circumference of his or her head and the total time spent doing homework in the last week. Plot the data on a scatter diagram. Do you think there is a hint of correlation present?

Line of best fit

If your scatter diagram leads you to suspect that there is linear correlation between the two variables plotted then you may reasonably try to draw a line of best fit. A simple, but not very accurate, way to do this is as follows.

(a) Calculate and plot the point $(\bar{x}, \bar{y})$ where $\bar{x}$ is the mean of the horizontal axis variable and $\bar{y}$ is the mean of the vertical axis variable.

(b) Draw a straight line which passes through $(\bar{x}, \bar{y})$ and which roughly leaves the same number of points of the scatter diagram above and below it, as in figure 4.8.

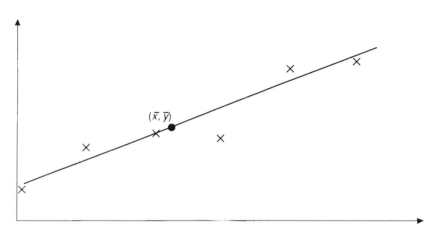

Figure 4.8

In some situations this rough and ready method will be adequate, particularly where it is easy to see where the line should go. In others cases, you may find it hard to judge where to place the line and may well want to draw the line accurately anyway. The method for doing this is described later in this chapter on pages 125–126.

EXERCISE 4A

For each of the sets of data below, draw a scatter diagram and comment on whether there appears to be any correlation. If there is then draw a possible line of best fit.

1 The mathematics and physics test results of 14 pupils.

Mathematics	15	23	78	91	46	27	41	62	34	17	77	49	55	71
Physics	62	36	92	70	67	39	61	40	55	33	65	59	35	40

2 The wine consumption in a country in millions of litres and the years 1993 to 2000.

Year	1993	1994	1995	1996	1997	1998	1999	2000
Consumption ($\times 10^6$ litres)	35.5	37.7	41.5	46.4	44.8	45.8	53.9	62.0

3 The number of hours of sunshine and the monthly rainfall, in centimetres, in an eight-month period.

	Jan	Feb	Mar	Apr	May	Jun	Jul	Aug
Sunshine (hours)	90	96	105	110	113	120	131	124
Rainfall (cm)	5.1	4.6	6.3	5.1	3.3	2.8	4.5	4.0

4 The annual salary, in thousands of pounds, and the average number of hours worked per week by seven people chosen at random.

Salary (× £1000)	5	7	13	14	16	20	48
Hours worked per week	18	22	35	38	36	36	32

5 The mean temperature in degrees centigrade and the amount of ice-cream sold in a supermarket in hundreds of litres.

	Apr	May	Jun	Jul	Aug	Sep	Oct	Nov
Mean temperature (°C)	9	13	14	17	16	15	13	11
Ice-cream sold (100 litres)	11	15	17	20	22	17	8	7

6 The reaction times of ten women of various ages.

Reaction time ($\times 10^{-3}$ s)	156	165	149	180	189	207	208	178
Age (years)	36	40	27	50	49	53	55	27

 In each of the questions in exercise 4A there are two variables. Which of the following words can be used to describe them: dependent, independent, random, non-random, controlled?

Product moment correlation

The scatter diagram in figure 4.1 revealed that there may be a mutual association between the goals scored and points totals of teams in the Premiership at the end of the 1998–99 season. In this section we set out to quantify this relationship.

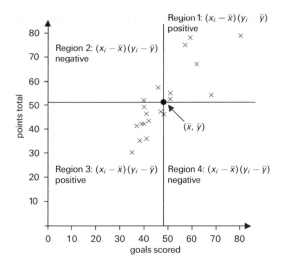

Figure 4.9

Figure 4.9 shows the scatter diagram of the last section again.

The mean number of goals scored is

$$x = \frac{\sum\limits_{i=1}^{n} x_i}{n} = \frac{80 + 59 + 57 + \cdots}{20} = \frac{959}{20} = 47.95$$

The mean points total is

$$y = \frac{\sum\limits_{i=1}^{n} y_i}{n} = \frac{79 + 78 + 75 + \cdots}{20} = \frac{1025}{20} = 51.25$$

[(x_i, y_i) are the various data points, for example (80, 79) for Manchester United. n is the number of such points, in this case 20, one for each club in the Premier League.]

You will see that the point $(\bar{x}, y)$ has also been plotted on the scatter diagram and lines drawn through this point parallel to the axes. These lines divide the scatter diagram into four regions.

You can think of the point $(\bar{x}, \bar{y})$ as the middle of the scatter diagram and so treat it as a new origin. Relative to $(\bar{x}, \bar{y})$, the co-ordinates of the various points are all of the form $(x_i - \bar{x}, y_i - \bar{y})$.

In regions 1 and 3 the product $(x_i - \bar{x})(y_i - \bar{y})$ is positive for every point.

In regions 2 and 4 the product $(x_i - \bar{x})(y_i - \bar{y})$ is negative for every point.

When there is positive correlation most or all of the data points will fall in regions 1 and 3 and so you would expect the sum of these terms to be positive and large. This sum is denoted by S_{xy}

$$S_{xy} = \sum_{i=1}^{n} (x_i - \bar{x})(y_i - \bar{y})$$

When there is negative correlation most or all of the points will be in regions 2 and 4 and so you would expect the sum of these terms (in the equation above) to be negative and large.

When there is little or no correlation the points will be scattered round all four regions. Those in regions 1 and 3 will result in positive values of $(x_i - \bar{x})(y_i - \bar{y})$ but when you add these to the negative values from the points in regions 2 and 4 you would expect most of them to cancel each other out. Consequently the total value of all the terms should be small.

By itself the actual value of S_{xy} does not tell you very much because:

1. no allowance has been made for the number of items of data;
2. no allowance has been made for the spread within the data;
3. no allowance has been made for the units of x and y.

Taking the first point into account converts S_{xy} into *sample covariance*. If allowance is made for the second and third points as well, the *product moment correlation coefficient* of the sample is found.

Covariance

The first point is met by finding the average value of S_{xy}. This is called the *sample covariance* and given by

$$\text{Sample covariance} = \frac{1}{n} S_{xy} = \frac{1}{n} \sum_{i} (x_i - \bar{x})(y_i - \bar{y})$$

where n is the number of data points.

Sample covariance may also be written in the form $\frac{1}{n} \sum_{i} (x_i y_i) - \bar{x}\bar{y}$.

This allows for the number of data points but not for the spread of the variables.

You can show that the two forms are equivalent, as follows.

$$\frac{1}{n} \sum_i (x_i - \bar{x})(y_i - \bar{y})$$

$$= \frac{1}{n} \sum_i (x_i y_i - x_i \bar{y} - \bar{x} y_i + \bar{x}\bar{y})$$

$$= \frac{1}{n} \sum_i x_i y_i - \bar{y} \times \frac{1}{n} \sum_i x_i - \bar{x} \times \frac{1}{n} \sum_i y_i + \frac{1}{n} \sum_i \bar{x}\bar{y}$$

$$\left\{ \frac{1}{n} \sum_i x_i = \bar{x} \right\} \quad \left\{ \frac{1}{n} \sum_i y_i = \bar{y} \right\}$$

$$= \frac{1}{n} \sum_i x_i y_i - \bar{y}\bar{x} - \bar{x}\bar{y} + \frac{1}{n} \times n\bar{x}\bar{y} \longleftarrow \left\{ \sum_i \bar{x}\bar{y} = n\bar{x}\bar{y} \right\}$$

$$= \frac{1}{n} \sum_i x_i y_i - \bar{x}\bar{y}$$

Pearson's product moment correlation coefficient

To allow for the spread of the data, you must standardise the values of $(x_i - \bar{x})$ and $(y_i - \bar{y})$.

A measure of the total square spread of the x values is given by

$$S_{xx} = \sum_i (x_i - \bar{x})^2$$

The average of this is

$$\frac{S_{xx}}{n} = \frac{1}{n} \sum_i (x_i - \bar{x})^2$$

Taking the square root of this gives a measure of the spread of the x values.

$$\sqrt{\frac{S_{xx}}{n}} = \sqrt{\frac{1}{n} \sum_i (x_i - \bar{x})^2}$$

This is the standard deviation of x calculated with divisor n. It can be obtained from a scientific calculator.

The corresponding measure of spread for the y values is

$$\sqrt{\frac{S_{yy}}{n}} = \sqrt{\frac{1}{n} \sum_i (y_i - \bar{y})^2}$$

Dividing by these gives a measure of linear correlation called *Pearson's product moment correlation coefficient*. It is denoted by the letter *r*.

$$r = \frac{\dfrac{S_{xy}}{n}}{\sqrt{\dfrac{S_{xx}}{n}}\sqrt{\dfrac{S_{yy}}{n}}}$$

$$= \frac{S_{xy}}{\sqrt{S_{xx}S_{yy}}} \quad \longleftarrow \quad \text{The } ns \text{ cancel out}$$

This can also be written as $r = \dfrac{\sum (x_i - \bar{x})(y_i - \bar{y})}{\sqrt{\sum (x_i - \bar{x})^2 \sum (y_i - \bar{y})^2}}$

An alternative form is $r = \dfrac{\dfrac{1}{n}\sum\limits_{i} x_i y_i - \bar{x}\bar{y}}{\sqrt{\left(\dfrac{1}{n}\sum\limits_{i} x_i^2 - \bar{x}^2\right)\left(\dfrac{1}{n}\sum\limits_{i} y_i^2 - \bar{y}^2\right)}}$

r is often just called the *correlation coefficient*.

The quantity, *r*, provides a standardised measure of correlation. Its value always lies within the range -1 to $+1$. (If you calculate a value outside this range, you have made a mistake.) A value of $+1$ means perfect positive correlation; in this case all the points on a scatter diagram would lie exactly on a straight line with positive gradient. Similarly a value of -1 means perfect negative correlation. These two cases are illustrated in figure 4.10.

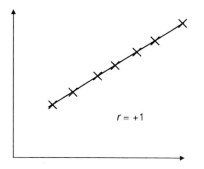

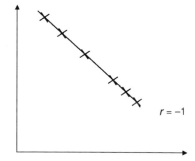

Figure 4.10 (i) *Perfect positive correlation* (ii) *Perfect negative correlation*

In cases of little or no correlation *r* takes values close to zero. The nearer the value of *r* is to $+1$ or -1, the stronger the correlation.

The calculation of sample covariance and Pearson's product moment correlation coefficient is shown in example 4.1, which is worked twice using the alternative formulae.

Be careful not to confuse the quantities denoted by S_{xx}, S_{yy} and S_{xy} with those denoted by s_{xx}, s_{yy} and s_{xy} in the earlier edition of this book and by some other authors. S_{xx} and S_{yy} are the sums of the squares and are not divided by n. Simlarly, S_{xy} is the sum of the terms $(x_i - \bar{x})(y_i - \bar{y})$ and is not divided by n.

Historical note

Karl Pearson was one of the founders of modern statistics. Born in 1857, he was a man of varied interests and practised law for three years before being appointed Professor of Applied Mathematics and Mechanics at University College, London in 1884. Pearson made contributions to various branches of mathematics but is particularly remembered for his work on the application of statistics to biological problems in heredity and evolution. He died in 1936.

EXAMPLE 4.1

A gardener wishes to know if plants which produce only a few potatoes produce larger ones. He selects five plants at random, sieves out the small potatoes, counts those remaining and weighs the largest one.

Number of potatoes, x	5	5	7	8	10
Weight of largest, y (grams)	240	232	227	222	215

Calculate the correlation coefficient between x and y and comment on the result.

SOLUTION (METHOD 1)

$n = 5$, $\bar{x} = 7$, $\bar{y} = 227.2$

x_i	y_i	$x_i - \bar{x}$	$y_i - \bar{y}$	$(x_i - \bar{x})^2$	$(y_i - \bar{y})^2$	$(x_i - \bar{x})(y_i - \bar{y})$
5	240	−2	12.8	4	163.84	−25.6
5	232	−2	4.8	4	23.04	−9.6
7	227	0	−0.2	0	0.04	0
8	222	1	−5.2	1	27.04	−5.2
10	215	3	−12.2	9	148.84	−36.6
Σ 35	1136	0	0	18	362.80	−77.0

$$\bar{x} = \frac{\Sigma x_i}{n} = \frac{35}{5} = 7 \qquad \bar{y} = \frac{\Sigma y_i}{n} = \frac{1136}{5} = 227.2$$

$$S_{xx} = \sum_i (x_i - \bar{x})^2 = 18$$

$$S_{yy} = \sum_i (y_i - \bar{y})^2 = 362.8$$

$$S_{xy} = \sum_i (x_i - \bar{x})(y_i - \bar{y}) = -77$$

$$\text{Correlation coefficient, } r = \frac{S_{xy}}{\sqrt{S_{xx}S_{yy}}} = \frac{-77}{\sqrt{18 \times 362.8}} = -0.953 \text{ (3 sf)}$$

SOLUTION (METHOD 2)

$n = 5$

x_i	y_i	x_i^2	y_i^2	$x_i y_i$
5	240	25	57 600	1200
5	232	25	53 824	1160
7	227	49	51 529	1589
8	222	64	49 284	1776
10	215	100	46 225	2150
Σ 35	1136	263	258 462	7875

$$\bar{x} = \frac{\Sigma x_i}{n} = \frac{35}{5} = 7 \qquad \bar{y} = \frac{\Sigma y_i}{n} = \frac{1136}{5} = 227.2$$

$$\frac{1}{n}\Sigma x_i^2 - \bar{x}^2 = \frac{263}{5} - 7^2 = 3.6$$

$$\frac{1}{n}\Sigma y_i^2 - \bar{y}^2 = \frac{258\,462}{5} - 227.2^2 = 72.56$$

$$Sample\ covariance = \frac{1}{n}\Sigma x_i y_i - \bar{x}\bar{y} = \frac{7875}{5} - 7 \times 227.2 = -15.4$$

$$Correlation\ coefficient,\ r = \frac{\left(\dfrac{1}{n}\underset{i}{\Sigma}\,x_i y_i - \bar{x}\bar{y}\right)}{\sqrt{\left(\dfrac{1}{n}\underset{i}{\Sigma}\,x_i^2 - \bar{x}^2\right)\left(\dfrac{1}{n}\underset{i}{\Sigma}\,y_i^2 - \bar{y}^2\right)}}$$

$$= \frac{-15.4}{\sqrt{3.6 \times 72.56}} = -0.953\ (3\ sf)$$

There is very strong negative linear correlation between the variables.

Large potatoes seem to be associated with small crop sizes.

1

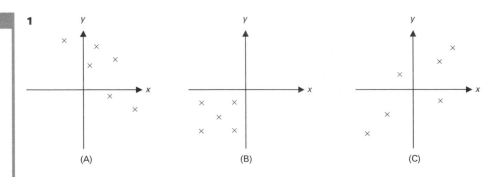

(A)　　　　　(B)　　　　　(C)

Three sets of bivariate data have been plotted on scatter diagrams, as illustrated. In each diagram the product moment correlation coefficient takes one of the values -1, -0.8, 0, 0.8, 1. State the appropriate value of the correlation coefficient corresponding to the scatter diagrams (A), (B) and (C).

[Cambridge]

In questions 2–6 find the sample covariance and Pearson's product moment correlation coefficient, r, for a number of bivariate samples.

The purpose of doing this on paper is to familiarise yourself with the routines involved. Most statisticians would actually do such calculations using a computer package, spreadsheet or a good calculator, and you should learn how to do this as well.

2

x	2	6	7	10
y	13	8	9	6

3

x	10	11	12	13	14	15	16	17
y	19	16	28	20	31	19	32	35

4

x	0	1	4	3	2
y	11	8	5	4	7

5

x	12	14	14	15	16	17	17	19
y	86	90	78	71	77	69	80	73

6

x	56	78	14	80	34	78	23	61
y	45	34	67	70	42	18	25	50

The meaning of a correlation coefficient

You have already seen that if the value of the correlation coefficient, r, is close to $+1$ or -1, you can be satisfied that there is linear correlation, and that if r is close to 0 there is probably little or no correlation. What happens in a case such as $r = 0.6$?

To answer this question you have to understand what r is actually measuring. The data which you use when calculating r are actually a *sample* of a parent bivariate distribution. You have only taken a few out of a very large number of points which could, in theory, be plotted on a scatter diagram such as figure 4.11. Each point (x_i, y_i) represents one possible value x_i of the variable X and one possible value y_i of Y.

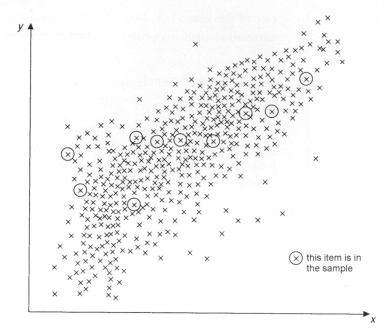

Figure 4.11 *Scatter diagram showing a sample from a large bivariate population*

There will be a level of correlation within the parent population and this is denoted by ρ (the Greek letter *rho*, pronounced 'row', as in 'row a boat').

Your calculated value of r, which is based on your sample points, can be used as an estimate for ρ. It can also be used to carry out a hypothesis test on the value of ρ, the parent population correlation coefficient. Used in this way it is a *test statistic.*

The simplest hypothesis test which you can carry out is that there is no correlation within the parent population. This gives rise to a null hypothesis:

H_0: $\rho = 0$ There is correlation between the two variables.

There are three possible alternative hypotheses, according to the sense of the situation you are investigating. These are:

1. H_1: $\rho \neq 0$ There is correlation between the variables.
   (two-tail test)

2. H_1: $\rho > 0$ There is positive correlation between the variables.
   (one-tail test)

3. H_1: $\rho < 0$ There is negative correlation between the variables.
   (one-tail test)

The test is carried out by comparing your value for r with the appropriate entry in a table of critical values. This will depend on the size of your sample, the significance level at which you are testing and whether your test is one- or two-tailed.

EXAMPLE 4.2

THE AVONFORD STAR

Letters to the Editor

Dear Sir,

The trouble with young people these days is that they watch too much TV. They just sit there gawping and become steadily less intelligent. I challenge you to carry out a proper test and I am sure you will find that the more TV children watch the less intelligent they are.

Yours truly,

Outraged senior citizen.

The editor of the *Avonford Star* was interested in the writer's point of view and managed to collect these data for the IQs of six children, and the number of hours of TV they watched in the previous week.

Hours of TV, x	9	11	14	7	10	9
IQ, y	142	112	100	126	109	88

The relevant hypothesis test for this situation would be

H_0: $\rho = 0$ There is no correlation between IQ and watching TV.

H_1: $\rho < 0$ There is negative correlation between IQ and watching TV. (one-tail test)

The editor decided to use the 5% significance level.

The critical value for $n = 6$ at the 5% significance level for a one-tail test is found from tables to be 0.7293.

	5%	$2\frac{1}{2}$%	1%	$\frac{1}{2}$%	1-Tail Test
	10%	5%	2%	1%	2-Tail Test
n					
1	–	–	–	–	
2	–	–	–	–	
3	0.9877	0.9969	0.9995	0.9999	
4	0.9000	0.9500	0.9800	0.9900	
5	0.8054	0.8783	0.9343	0.9587	
6	0.7293	0.8114	0.8822	0.9172	
7	0.6694	0.7545	0.8329	0.8745	
8	0.6215	0.7067	0.7887	0.8343	
9	0.5822	0.6664	0.7498	0.7977	
10	0.5494	0.6319	0.7155	0.7646	
11	0.5214	0.6021	0.6851	0.7348	
12	0.4973	0.5760	0.6581	0.7079	
13	0.4762	0.5529	0.6339	0.6835	
14	0.4575	0.5324	0.6120	0.6614	
15	0.4409	0.5140	0.5923	0.6411	

Figure 4.12 *Extract from table of values for the product moment correlation coefficient, r*

The calculation of the correlation coefficient can be set out as follows. (Notice that most of the figures are given to two decimal places, but greater accuracy is used in the actual calculation.)

x_i	y_i	x_i^2	y_i^2	$x_i y_i$
9	142	81	20 164	1278
11	112	121	12 544	1232
14	100	196	10 000	1400
7	126	49	15 876	882
10	109	100	11 881	1090
9	88	81	7 744	792
Σ 60	677	628	78 209	6674

$$\bar{x} = \frac{\Sigma x_i}{n} = \frac{60}{6} = 10 \qquad \bar{y} = \frac{\Sigma y_i}{n} = \frac{677}{6} = 112.83$$

$$\frac{1}{n} \sum_i x_i^2 - \bar{x}^2 = \frac{628}{6} - 10^2 = 4.66\ldots$$

$$\frac{1}{n} \sum_i y_i^2 - \bar{y}^2 = \frac{78\,209}{6} - 112.83^2 = 303.4\ldots$$

$$\frac{1}{n} \Sigma x_i y_i - \bar{x}\bar{y} = \frac{6674}{6} - 10 \times 112.83 = -16.00$$

Correlation coefficent, $r = \dfrac{\dfrac{1}{n} \sum_i x_i y_i - \bar{x}\bar{y}}{\sqrt{\left(\dfrac{1}{n} \sum_i x_i^2 - \bar{x}^2\right)\left(\dfrac{1}{n} \sum_i y_i^2 - \bar{y}^2\right)}}$

$$= \frac{-16.00}{\sqrt{4.66\ldots \times 303.4\ldots}} = -0.43$$

Since $0.43 < 0.7293$, the critical value, the null hypothesis is accepted.

The evidence from this small sample is not sufficient to justify the claim that there is negative correlation between IQ and TV watching. Notice however that, even if it had been, this would not necessarily have supported the Outraged senior citizen's claim that TV lowers IQ. It could just be that people with higher IQs watch less TV.

 This sort of test assumes that both variables are random. For large data sets this is usually the case if the scatter diagram gives an approximately elliptical distribution.

The extract from the tables gives the critical value of r for various values of the significance level and the sample size, n. You will, however, find some tables where n is replaced by ν, the *degrees of freedom*. (ν is the Greek letter n and is pronounced 'new'.) Degrees of freedom are covered in the next section.

Degrees of freedom

Here is an example where you have just two data points.

	Sean	Iain
Height of an adult man (m)	1.70	1.90
The mortgage on his house (£)	15 000	45 000

When you plot these two points on a scatter diagram it is possible to join them with a perfect straight line and you might be tempted to conclude that taller men have larger mortgages on their houses.

This conclusion would clearly be wrong. It is based on the data from only two men so you are bound to be able to join the points on the scatter diagram with a straight line and calculate r to be either $+1$ or -1 (providing their heights and/or mortgages are not the same). In order to start to carry out a test you need the data for a third man, say Dafyd (height 1.75 m and mortgage £37 000). When his data are plotted on the scatter diagram in figure 4.13 you can see how close it lies to the line between Sean and Iain.

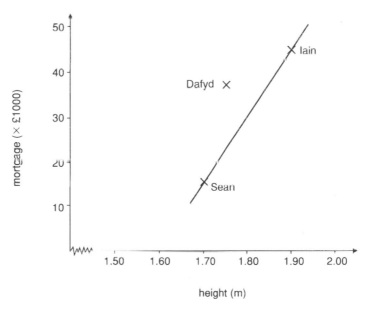

Figure 4.13

So the first two data points do not count towards a test for linear correlation. The first one to count is point number three. Similarly if you have n points, only $n - 2$ of them count towards any test. $n - 2$ is called the *degrees of freedom* and denoted by ν. It is the number of free variables in the system. In this case it is the number of points, n, less the 2 that have effectively been used to define the line of best fit.

In the case of the three men with their mortgages you would actually draw a line of best fit through all three, rather than join any particular two. So you cannot say that any two particular points have been taken out to draw the line of best fit, merely that the system as a whole has lost two.

Tables of critical values of correlation coefficients can be used without understanding the idea of degrees of freedom, but the idea is an important one which you will often use as you learn more statistics. In general

$$\text{degrees of freedom} = \text{sample size} - \text{number of restrictions.}$$

Different types of restriction apply in different statistical procedures.

Interpreting correlation

You need to be on your guard against drawing spurious conclusions from significant correlation coefficients.

Correlation does not imply causation

Figures for the years 1985–93 show a high correlation between the sales of personal computers and those of microwave ovens. There is of course no direct connection between the two variables. You would be quite wrong to conclude that buying a microwave oven predisposes you to buy a computer as well, or vice versa.

Although there may be a high level of correlation between variables A and B it does not mean that A causes B or that B causes A. It may well be that a third variable C causes both A and B, or it may be that there is a more complicated set of relationships. In the case of personal computers and microwaves, both are clearly caused by the advance of modern technology.

Non-linear correlation

A low value of r tells you that there is little or no *linear* correlation. There are other forms of correlation as illustrated in figure 4.14.

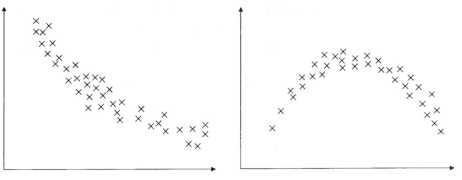

Figure 4.14 *Scatter diagrams showing non-linear correlation*

These diagrams show that there is an association between the variables, but not one of linear correlation.

Extrapolation

A linear relationship established over a particular domain should not be assumed to hold outside this range. For instance, there is strong correlation between the age in years and the 100-metre times of female athletes between the ages of 10 and 20 years. To extend the connection, as shown in figure 4.15, would suggest that veteran athletes are quicker than athletes who are in their prime and, if they live long enough, can even run 100 metres in no time at all!

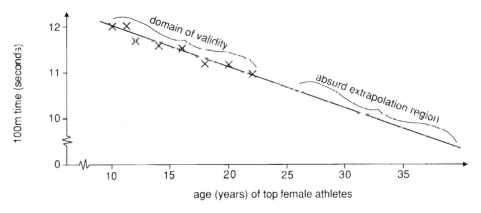

Figure 4.15

Most of the questions in this exercise involve unrealistically small samples. They are meant to help you understand the principles involved in testing for correlation. When you come to do such testing on real data, you would hope to be able to use much larger samples.

1 A language teacher wished to test whether there is any correlation between students' ability in their own and a foreign language. Accordingly she collects the marks of eight students, all native English speakers, in their end of year examinations in English and French.

Candidate	A	B	C	D	E	F	G	H
English	65	34	48	72	58	63	26	80
French	74	49	45	80	63	72	12	75

(i) Calculate the product moment correlation coefficient.

(ii) State the null and alternative hypotheses.

(iii) Using the correlation coefficient as a test statistic, carry out the test at the 5% significance level.

2 'You can't win without scoring goals.' So says the coach of a netball team. Jamila, who believes in solid defensive play, disagrees and sets out to prove that there is no correlation between scoring goals and winning matches. She collects the following data for the goals scored and the points gained by 12 teams in a netball league:

Goals scored, x	41	50	54	47	47	49	52	61	50	29	47	35
Points gained, y	21	20	19	18	16	14	12	11	11	7	5	2

(i) Calculate the product moment correlation coefficient.

(ii) State suitable null and alternative hypotheses, indicating whose position each represents.

(iii) Carry out the hypothesis test at the 5% significance level and comment on the result.

3 A sports reporter believes that those who are good at the high jump are also good at the long jump, and vice versa. He collects data on the best performances of nine athletes, as follows.

Athlete	A	B	C	D	E	F	G	H	I
High jump, x (metres)	2.0	2.1	1.8	2.1	1.8	1.9	1.6	1.8	1.8
Long jump, y (metres)	8.0	7.6	6.4	6.8	5.8	8.0	5.5	5.5	6.6

(i) Calculate the product moment correlation coefficient.

(ii) State suitable null and alternative hypotheses.

(iii) Carry out the hypothesis test at the 5% significance level and comment on the result.

4 It is widely believed that those who are good at chess are good at bridge, and vice versa. A commentator decides to test this theory using as data the grades of a random sample of eight people who play both games.

Player	A	B	C	D	E	F	G	H
Chess grade	160	187	129	162	149	151	189	158
Bridge grade	75	100	75	85	80	70	95	80

(i) Calculate the product moment correlation coefficient.

(ii) State suitable null and alternative hypotheses.

(iii) Carry out the hypothesis test at the 5% significance level. Do these data support this belief at this significance level?

5 A biologist believes that a particular type of fish develops black spots on its scales in water that is polluted by certain agricultural fertilisers. She catches a number of fish; for each one she counts the number of black spots on its scales and measures the concentration of the pollutant in the water it was swimming in. She uses these data to test for positive linear correlation between the number of spots and the level of pollution.

Fish	A	B	C	D	E	F	G	H	I	J
Pollutant concentration (parts per million)	124	59	78	79	150	12	23	45	91	68
No. of black spots	15	8	7	8	14	0	4	5	8	8

(i) Calculate the product moment correlation coefficient.
(ii) State suitable null and alternative hypotheses.
(iii) Carry out the hypothesis test at the 2% significance level. What can the biologist conclude?

6 Andrew claims that the older you get, the slower is your reaction time. His mother disagrees, saying the two are unrelated. They decide that the only way to settle the discussion is to carry out a proper test. A few days later they are having a small party and so ask their twelve guests to take a test that measures their reaction times. The results are as follows:

Age	Reaction time (s)	Age	Reaction time (s)
78	0.8	35	0.5
72	0.6	30	0.3
60	0.7	28	0.4
56	0.5	20	0.4
41	0.5	19	0.3
39	0.4	10	0.3

Carry out the test at the 5% significance level, stating the null and alternative hypotheses. Who won the argument, Andrew or his mother?

7 The teachers at a school have a discussion as to whether girls in general run faster or slower as they get older. They decide to collect data for a random sample of girls the next time the school cross country race is held (which everybody has to take part in). They collect the following data, with the times given in minutes and the ages in years (the conversion from months to decimal parts of a year has already been carried out).

Age	Time	Age	Time	Age	Time
11.6	23.1	18.2	45.0	13.9	29.1
15.0	24.0	15.4	23.2	18.1	21.2
18.8	45.0	14.4	26.1	13.4	23.9
16.0	25.2	16.1	29.4	16.2	26.0
12.8	26.4	14.6	28.1	17.5	23.4
17.6	22.9	18.7	45.0	17.0	25.0
17.4	27.1	15.4	27.0	12.5	26.3
13.2	25.2	11.8	25.4	12.7	24.2
14.5	26.8				

(i) State suitable null and alternative hypotheses and decide on an appropriate significance level for the test.

(ii) Calculate the product moment correlation coefficient and state the conclusion from the test.

(iii) Plot the data on a scatter diagram and identify any outliers. Explain how they could have arisen.

(iv) Comment on the validity of the test.

8 The manager of a company wishes to evaluate the success of its training programme. One aspect which interests her is to see if there is any relationship between the amount of training given to employees and the length of time they stay with the company before moving on to jobs elsewhere. She does not want to waste company money training people who will shortly leave. At the same time she believes that the more training employees are given the longer they will stay. She collects data on the average number of days training given per year to 25 employees who have recently left for other jobs, and the length of time they worked for the company.

Training (days/year)	Work (days)	Training (days/year)	Work (days)	Training (days/year)	Work (days)
2.0	354	3.4	760	1.2	132
4.0	820	1.8	125	4.5	1365
0.1	78	0.0	28	1.0	52
5.6	1480	5.7	1360	7.8	1080
9.1	980	7.2	1520	3.7	508
2.6	902	7.5	1380	10.9	1281
0.0	134	3.0	121	3.8	945
2.6	252	2.8	457	2.9	692
7.2	867				

(i) Calculate the product moment correlation coefficient.

(ii) State suitable null and alternative hypotheses.

(iii) Carry out the hypothesis test at the 5% significance level.

(iv) Plot the data on a scatter diagram.

(v) What conclusions would you come to if you were the manager?

9 Charlotte is a campaigner for temperance, believing that drinking alchohol is an evil habit. Michel, representative of a wine company presents her with these figures which he claims show that wine drinking is good for marriages.

Country	Wine consumption (litres/person/year)	Divorce rate (per 1000 inhabitants)
Belgium	20	2.0
Denmark	20	2.7
Germany	26	2.2
Greece	33	0.6
Italy	63	0.4
Portugal	54	0.9
Spain	41	0.6
U.K.	13	2.9

(i) Write Michel's claim in the form of a hypothesis test and carry it out.

(ii) Charlotte claims that Michel is 'indulging in pseudo-statistics'. What arguments could she use to support this point of view?

10 The values of x and y in the table are the marks obtained in an intelligence test and a university examination respectively by 20 medical students. The data are plotted in the scatter diagram.

x	98	51	71	57	44	59	75	47	39	58
y	85	40	30	25	50	40	50	35	25	90
x	77	65	58	66	79	72	45	40	49	76
y	65	25	70	45	70	50	40	20	30	60

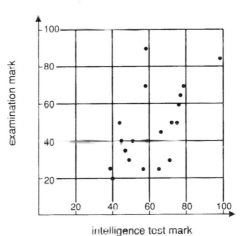

Given that $\Sigma x = 1226$, $\Sigma y = 945$, $\Sigma x^2 = 79\,732$, $\Sigma y^2 = 52\,575$ and $\Sigma xy = 61\,495$, calculate the product moment correlation coefficient r to 2 decimal places.

Referring to the evidence provided by the diagram and the value of r, comment briefly on the correlation between the two sets of marks.

Now eliminate from consideration those 10 students whose values of x are less than 50 or more than 75. Calculate the new value of r for the marks of the remaining students. What does the comparison with the earlier value of r seem to indicate?

[MEI]

11 The following data refer to the price (in £ per kg) of mammoth meat in London for the four quarters of the years 1994–99.

Year	Quarter	Price	Year	Quarter	Price
1999	4	£5.00	1996	4	£4.70
	3	£6.00		3	£5.70
	2	£6.90		2	£6.80
	1	£6.10		1	£5.80
1998	4	£4.90	1995	4	£4.50
	3	£5.80		3	£5.60
	2	£6.90		2	£6.50
	1	£5.80		1	£5.40
1997	4	£4.80	1994	4	£4.40
	3	£5.80		3	£5.50
	2	£6.80		2	£6.30
	1	£5.90		1	£5.20

(i) Form these data into a bivariate set by taking each price with the one before it: $(5.00, 6.00)$, $(6.00, 6.90)$, $(6.90, 6.10)$, ..., $(6.30, 5.20)$. Work out the correlation coefficient for this set and call it r_1.

(ii) Form another bivariate set by doing the same thing but with data from two quarters apart: $(5.00, 6.90)$, $(6.00, 6.10)$, ..., $(5.50, 5.20)$. Work out the correlation coefficient for this set and call it r_2.

(iii) Repeat this procedure for the set of bivariate data three quarters apart and find r_3.

(iv) Repeat this procedure for the set of bivariate data four quarters apart and find r_4.

(v) What do you notice? This technique of correlating time-dependent data with itself is called *autocorrelation* and is used to detect cyclic or seasonal variations in the data. What do the different values of r tell you about these data?

12 Examine this newspaper report critically.

British publishers hold the key to continental exchange rates

Startling new evidence has come to light of the effect of the flourishing British publishing industry on the economies of France and Germany. The figures for UK Book Title Production and for the Franc/Mark exchange rate (adjusted for differentials in producer prices) are given in the table for the period 1980–90.

The correlation between the two is significant even at the 1% level, showing beyond all reasonable doubt that our publishers are steadily undermining the German currency.

Year	Books	Exch. rate
1980	48 158	96
1981	43 083	93
1982	48 307	95
1983	51 071	100
1984	51 555	101
1985	52 994	97
1986	52 496	99
1987	54 746	105
1988	56 514	105
1989	61 361	100
1990	63 756	104

(Note: The article is fictitious but the figures are real.)

Rank correlation

THE AVONFORD STAR

Punch-up at village fete

Pandemonium broke out at the Normanton village fete last Saturday when the adjudication for the Tomato of the Year competition was announced. The two judges completely failed to agree in their rankings and so a compromise winner was chosen to the fury of everybody (except the winner).

Following the announcement there was a moment of stunned silence, followed by shouts of 'Rubbish', 'It's a fix', 'Go home' and further abuse. Then the tomatoes started to fly and before long fighting broke out.

By the time the police arrived on the scene ten people were injured, including last year's winner Bert Wallis who lost three teeth in the scrap. Both judges had escaped unhurt.

Angry Bert Wallis, nursing a badly bruised jaw, said 'The competition was a nonsense. The judges were useless. Their failure to agree shows they did not know what they were looking for'. But fete organiser Margaret Bramble said this was untrue. 'The competition was completely fair; both judges know a good tomato when they see one,' she claimed.

The judgement that caused all the trouble was as follows:

Tomato	A	B	C	D	E	F	G	H
Judge 1	1	8	4	6	2	5	7	3
Judge 2	7	2	3	4	6	8	1	5
Total	8	10	⑦	10	8	13	8	8
			Winner					

You will see that both judges ranked the eight entrants, 1st, 2nd, 3rd, ..., 8th. The winner, C, was placed 4th by one judge and 3rd by the other. Their rankings look different so perhaps they were using different criteria on which to assess them. How can you use these data to decide whether that was or was not the case?

One way would be to calculate a correlation coefficient and use it to carry out a hypothesis test:

H_0: $\rho = 0$ There is no correlation.
H_1: $\rho > 0$ There is positive correlation.

The null hypothesis, H_0, represents something like Bert Wallis's view, the alternative hypothesis that of Margaret Bramble.

However, the data you have are of a different type from any that you have used before for calculating correlation coefficients. In the point $(1, 7)$, corresponding to tomato A, the numbers 1 and 7 are *ranks* and not scores (like marks in an examination or measurements). It is, however, possible to calculate a *rank correlation coefficient*, and in the same way as before.

Tomato	Judge 1	Judge 2			
	x_i	y_i	x_i^2	y_i^2	$x_i y_i$
A	1	7	1	49	7
B	8	2	64	4	16
C	4	3	16	9	12
D	6	4	36	16	24
E	2	6	4	36	12
F	5	8	25	64	40
G	7	1	49	1	7
H	3	5	9	25	15
Total Σ	36	36	204	204	133

$$\bar{x} = \frac{\Sigma x_i}{n} = \frac{36}{8} = 4.5 \qquad \text{Similarly } \bar{y} = \frac{36}{8} = 4.5$$

$$\frac{1}{n} \sum_i x_i^2 - \bar{x}^2 = \frac{204}{8} - 4.5^2 = 5.25$$

$$\text{Similarly } \frac{1}{n} \sum_i y_i^2 - \bar{y}^2 = 5.25$$

$$\frac{1}{n} \Sigma x_i y_i - \bar{x}\bar{y} = \frac{133}{8} - 4.5 \times 4.5 = -3.625$$

$$\text{Correlation coefficient, } r = \frac{-3.625}{\sqrt{5.25 \times 5.25}} = -0.69$$

Since the correlation coefficient is negative there can be no possibility of accepting H_1, therefore you accept H_0. There is no evidence of agreement between the judges. Remember this was a one-tail test for positive correlation.

Spearman's coefficient of rank correlation

The calculation in the previous example is usually carried out by a quite different, but equivalent, procedure and the resulting correlation coefficient is called *Spearman's coefficient of rank correlation* and denoted by r_s.

The procedure is summarised by the formula

$$r_s = 1 - \frac{6\Sigma d_i^2}{n(n^2 - 1)}$$

where d_i is the difference in the ranks given to the ith item.

The calculation is then as follows:

Tomato	Judge 1 x_i	Judge 2 y_i	$d_i = x_i - y_i$	d_i^2
A	1	7	−6	36
B	8	2	6	36
C	4	3	1	1
D	6	4	2	4
E	2	6	−4	16
F	5	8	−3	9
G	7	1	6	36
H	3	5	−2	4
Total				$\Sigma d_i^2 = 142$

$$r_s = 1 - \frac{6\Sigma d_i^2}{n(n^2 - 1)} = 1 - \frac{6 \times 142}{8(8^2 - 1)}$$
$$= 1 - 1.690$$
$$= -0.690$$

You will see that this is the same answer as before, but the working is much shorter. It is not difficult to prove that the two methods are equivalent and this is done on pages 137–138 of the Appendix.

Critical values for Spearman's rank correlation coefficient, however, are different from those for Pearson's product moment correlation coefficient, so you must always be careful to use the appropriate tables.

The calculation is often carried out with the data across the page rather than in columns and this is shown in the next example.

EXAMPLE 4.3

During their course two trainee tennis coaches, Rachael and Leroy, were shown videos of seven people, A, B, C,..., G, doing a top spin service and were asked to rank them in order according to the quality of their style. They placed them as follows:

Rank order	1	2	3	4	5	6	7
Rachael	B	G	F	D	A	C	E
Leroy	F	B	D	E	G	A	C

(i) Find Spearman's coefficient of rank correlation.

(ii) Use it to test whether there is evidence at the 5% level of positive correlation between their judgements.

SOLUTION

(i) The rankings are:

Person	A	B	C	D	E	F	G	
Rachael	5	1	6	4	7	3	2	
Leroy	6	2	7	3	4	1	5	
d_i	-1	-1	-1	1	3	2	-3	$n = 7$
d_i^2	1	1	1	1	9	4	9	$\Sigma d_i^2 = 26$

$$r_s = 1 - \frac{6\Sigma d_i^2}{n(n^2 - 1)} = 1 - \frac{6 \times 26}{7(7^2 - 1)}$$

$$= 0.54 \quad (2\ dp)$$

(ii) H_0: $\rho = 0$ There is no correlation between their rankings.

H_1: $\rho > 0$ There is positive correlation between their rankings.

Significance level 5%, one-tail test.

From tables the critical value of r_s for a one-tail test at this significance level for $n = 7$ is 0.7143.

$0.54 < 0.7143$ so H_0 is accepted.

	5%	$2\frac{1}{2}\%$	1%	$\frac{1}{2}\%$	1-Tail Test
					2-Tail Test
	10%	5%	2%	1%	
n					
1	–	–	–	–	
2	–	–	–	–	
3	–	–	–	–	
4	1.0000	–	–	–	
5	0.9000	1.0000	1.0000	–	
6	0.8286	0.8857	0.9429	1.0000	
7	0.7143	0.7857	0.8929	0.9286	
8	0.6429	0.7381	0.8333	0.8810	
9	0.6000	0.7000	0.7833	0.8333	
10	0.5636	0.6485	0.7455	0.7939	

Figure 4.16 *Extract from table of critical values for Spearman's rank correlation coefficient, r_s*

There is insufficient evidence to claim positive correlation between Rachael's and Leroy's rankings.

Historical note

Charles Spearman was born in London in 1863. After serving 14 years in the army as a cavalry officer, he went to Leipzig to study psychology. On completing his doctorate there he became a lecturer, and soon afterwards a professor, at University College, London. He pioneered the application of statistical techniques within psychology and developed the technique known as factor analysis in order to analyse different aspects of human ability. He died in 1945.

Tied ranks

If several items are ranked equally you give them the mean of the ranks they would have had if they had been slightly different from each other.

For example A, B, ..., J are ranked

1	2=	2=	4	5	6=	6=	6=	9	10
C	G	J	A	D	B	I	F	E	H

G and J are both 2= and so are given the rank $\dfrac{2+3}{2} = 2.5$.

B, I and F are all 6= and so are given the rank $\dfrac{6+7+8}{3} = 7$.

When to use rank correlation

Sometimes your data will be available in two forms, as values of variables or in rank order. If you have the choice you will usually work out the correlation coefficient from the variable values rather than the ranks.

If may well be the case, however, that only ranked data are available to you and in that case you have no choice but to use them. It may also be that while you could collect variable values as well, it would not be worth the time, trouble or expense to do so.

Pearson's product moment correlation coefficient is a measure of *linear* correlation and so is not appropriate for non-linear data like those illustrated in the scatter diagram in figure 4.17. You may, however, use rank correlation to investigate whether one variable generally increases (or decreases) as the other increases.

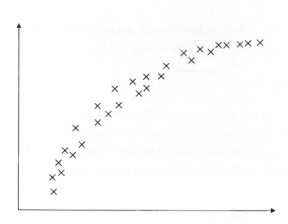

Figure 4.17 *Non-linear data with a high degree of rank correlation*

You should, however, always look at the sense of your data before deciding which is the more appropriate form of correlation to use.

Note

Spearman's rank correlation coefficient provides one among many statistical tests that can be carried out on ranks rather than variable values. Such tests are examples of *non-parametric tests*. A non-parametric test is a test on some aspect of a distribution which is not specified by its defining parameters.

EXERCISE 4D

1 Two independent assessors awarded marks to each of five projects. The results were as shown in the table.

Project	A	B	C	D	E
First assessor	38	91	62	83	61
Second assessor	56	84	41	85	62

Calculate Spearman's rank correlation coefficient for these data.

[Cambridge]

2 The order of merit of ten individuals at the start and finish of a training course were:

Individual	A	B	C	D	E	F	G	H	I	J
Order at start	1	2	3	4	5	6	7	8	9	10
Order at finish	5	3	1	9	2	6	4	7	10	8

Find Spearman's coefficient of rank correlation between the two orders.

3 Two people, *X* and *Y*, were asked to give marks out of 20 for seven brands of fish fingers. The results are recorded in the table.

Brand	A	B	C	D	E	F	G
X's mark	8	10	18	2	1	4	15
Y's mark	5	14	12	9	4	1	19

Construct a table of ranks and calculate Spearman's rank correlation coefficient.

[Cambridge]

4 In a marrow competition there were seven entrants and two judges who made the following decisions:

Marrow	A	B	C	D	E	F	G
Judge X	6	1	7	4	2	5	3
Judge Y	1	6	2	5	7	4	3

(i) Calculate Spearman's coefficient of rank correlation between the judges.
(ii) Fred Giles says the judges are looking for opposite qualities. Do the figures support this view at the 5% significance level?

5 A psychologist obtained scores by nine university entrants in three tests (A, B and C). The scores in tests A and B were as follows:

Entrant	1	2	3	4	5	6	7	8	9
A score	8	3	9	10	4	9	6	4	5
B score	7	8	5	9	10	6	3	4	7

Calculate a coefficient of rank correlation between the two sets of scores. The coefficient obtained between the A and C scores was 0.71 and that between the B and C scores was 0.62. What advice would you give the psychologist if he wished to use fewer than three tests?

[Cambridge]

6 In a driving competition there were eight contestants and three judges who placed them in rank order as shown in the table below:

Competitor	A	B	C	D	E	F	G	H
Judge X	2	5	6	1	8	4	7	3
Judge Y	1	6	8	3	7	2	4	5
Judge Z	2–	2=	6=	4	6=	1	6=	5

(i) Which two judges agreed the most?
(ii) Stating suitable null and alternative hypotheses, carry out a hypothesis test on the level of agreement of these two judges.

7 In a skating competition two judges place the contestants, in descending order of merit, as follows:

Judge 1		C	E	D	F	A	G	I	J	B	H
Judge 2			F	G	D	A	I	C	H	E	J	B

  (i) Find Spearman's coefficient of rank correlation between the two orders.
  (ii) Stating suitable null and alternative hypotheses, test at the 5% significance level whether the judges were in broad agreement with each other.

8 A coach wanted to test his theory that, although athletes have specialisms, it is still true that those who run fast at one distance are also likely to run fast at another distance. He selected six athletes at random to take part in a test and invited them to compete over 100 m and over 1500 m.

The times and places of the six athletes were as follows:

Athlete	100 m time	100 m rank	1500 m time	1500 m rank
Allotey	9.8 s	1	3 m 42 s	1
Chell	10.9 s	6	4 m 11 s	2
Giles	10.4 s	2	4 m 19 s	6
Mason	10.5 s	3	4 m 18 s	5
O'Hara	10.7 s	5	4 m 12 s	3
Stuart	10.6 s	4	4 m 16 s	4

  (i) Calculate the Pearson product moment and Spearman's rank correlation coefficients for these data.
  (ii) State suitable null and alternative hypotheses and carry out hypothesis tests on these data.
  (iii) State which you consider to be the more appropriate correlation coefficient in this situation, giving your reasons.

9 At the end of a word-processing course the trainees are given a document to type. They are assessed on the time taken and on the quality of their work. For a random sample of 12 trainees the following results were obtained.

Trainee	A	B	C	D	E	F	G	H	I	J	K	L
Quality (%)	97	96	94	91	90	87	86	83	82	80	77	71
Time (s)	210	230	198	204	213	206	200	186	192	202	191	199

  (i) Calculate Spearman's coefficient of rank correlation for the data. Explain what the sign of your correlation coefficient indicates about the data.
  (ii) Carry out a test, at the 5% level of significance, of whether or not there is any correlation between time taken and quality of work for trainees who have attended this word-processing course. State clearly the null and alternative hypotheses under test and the conclusion reached.

[MEI]

10 A school holds an election for parent governors. Candidates are invited to write brief autobiographies and these are sent out at the same time as the voting papers.

After the election, one of the candidates, Mr Smith, says that the more words you wrote about yourself the more votes you got. He sets out to 'prove this statistically' by calculating the product moment correlation between the number of words and the number of votes.

Candidate	A	B	C	D	E	F	G
Number of words	70	101	106	232	150	102	98
Number of votes	99	108	97	144	94	54	87

(i) Calculate the product moment correlation coefficient.

Mr Smith claims that this proves his point at the 5% significance level.

(ii) State his null and alternative hypotheses and show how he came to his conclusion.

(iii) Calculate Spearman's rank correlation coefficient for these data.

(iv) Explain the difference in the two correlation coefficients and criticise the procedure Mr Smith used in coming to his conclusion.

11 To test the belief that milder winters are followed by warmer summers, meteorological records are obtained for a random sample of ten years. For each year the mean temperatures are found for January and July. The data, in degrees Celsius, are given below.

January	8.3	7.1	9.0	1.8	3.5	4.7	5.8	6.0	2.7	2.1
July	16.2	13.1	16.7	11.2	14.9	15.1	17.7	17.3	12.3	13.4

(i) Rank the data and calculate Spearman's rank correlation coefficient.

(ii) Test, at the 2.5% level of significance, the belief that milder winters are followed by warmer summers. State clearly the null and alternative hypotheses under test.

(iii) Would it be more appropriate, less appropriate or equally appropriate to use the product moment correlation coefficient to analyse these data? Briefly explain why.

[MEI]

12 The following data, referring to the ordering of perceived risk of 25 activities and technologies and actual fatality estimates, were obtained in a study in the United States. Use these data to test at the 5% significance level for positive correlation between:

(i) the League of Women Voters and college students

(ii) experts and actual fatality estimates

(iii) college students and experts.

Comment on your results and identify any outliers in the three sets of bivariate data you have just used.

	League of Women Voters	College students	Experts	Actual fatalaties (estimates)
Nuclear power	1	1	18	16
Motor vehicles	2	4	1	3
Handguns	3	2	4	4
Smoking	4	3	2	1
Motorcycles	5	5	6	6
Alcoholic beverages	6	6	3	2
General (private) aviation	7	12	11	11
Police work	8	7	15	18
Surgery	9	10	5	8
Fire fighting	10	9	16	17
Large construction	11	11	12	12
Hunting	12	15	20	14
Mountain climbing	13	17	24	21
Bicycles	14	19	13	13
Commercial aviation	15	13	14	20
Electric power (non-nuclear)	16	16	8	5
Swimming	17	25	9	7
Contraceptives	18	8	10	19
Skiing	19	20	25	24
X-rays	20	14	7	9
High school & college football	21	21	22	23
Railroads	22	18	17	10
Power mowers	23	23	23	22
Home appliances	24	22	19	15
Vaccinations	25	24	21	25

Source: Shwing and Albers, *Societal Risk Assessment*, Plenum

The least squares regression line

A correlation coefficient provides you with a measure of the level of association between the two variables in a bivariate distribution.

If this indicates that there is a relationship, your question will be 'What is it?' In the case of linear correlation it can be expressed algebraically as a linear equation or geometrically as a straight line on the scatter diagram.

At the start of this chapter you saw how to draw a line of best fit through a set of points on a scatter diagram by eye. You did this by drawing it through the mean point, leaving roughly the same number of points above and below it. This is obviously a very imprecise method.

Before you do any calculations you first need to look carefully at the two variables that give rise to your data. It is normal practice to plot the *dependent variable* on the vertical axis and the *independent variable* on the horizontal axis. In the example which follows, the independent variable is the time at which measurements are made. (Notice that this is a non-random variable.) The procedure leads to the equation of the *regression line*, the line of best fit in these circumstances.

Look at the scatter diagram (figure 4.18) showing the n points A (x_1, y_1), B $(x_2, y_2), \ldots,$ N (x_n, y_n). On it is marked a possible line of best fit l. If the line l passed through all points there would be no problem since there would be perfect linear correlation. It does not, of course, pass though all the points and you would be very surprised if such a line did in any real situation.

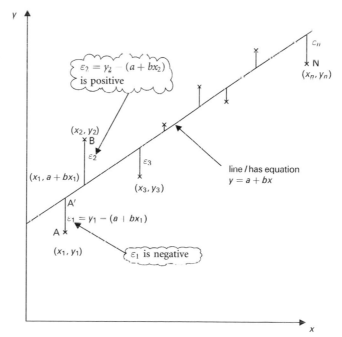

Figure 4.18 *Bivariate data plotted on a scatter diagram with the regression line, l, $y = a + bx$, and the residuals $\varepsilon_1, \varepsilon_2, \ldots, c_n$*

By how much is it missing the points? The answer to that question is shown by the vertical lines from the points to the line. Their lengths $\varepsilon_1, \varepsilon_2, \ldots, \varepsilon_n$ are called the *residuals* and represent the variation which is not explained by the line l. The *least squares regression line* is the line which produces the least possible value of the sum of the squares of the residuals, $\varepsilon_1^2 + \varepsilon_2^2 + \cdots + \varepsilon_n^2$.

If the equation of the line l is $y = a + bx$, then it is easy to see that the point A$'$ on the diagram, directly above A, has co-ordinates $(x_1, a + bx_1)$ and so the corresponding residual, ε_1, is given by $\varepsilon_1 = y_1 - (a + bx_1)$. Similarly for $\varepsilon_2, \varepsilon_3, \ldots, \varepsilon_n$.

The problem is to **find the values of the constants** a and b in the equation of the line l which make $\varepsilon_1^2 + \varepsilon_2^2 + \cdots + \varepsilon_n^2$ a minimum for any particular set of data, that is to minimise

$$[y_1 - (a + bx_1)]^2 + [y_2 - (a + bx_2)]^2 + \cdots + [y_n - (a + bx_n)]^2$$

The mathematics involved in doing this is not particularly difficult and is given on pages 139–141 of the Appendix. There are a number of equivalent ways of writing the resulting equation of the regression line.

$$y - \bar{y} = \frac{S_{xy}}{S_{xx}}(x - \bar{x})$$

$$\text{or} \quad y - \bar{y} = \frac{\sum_i (x_i - \bar{x})(y_i - \bar{y})}{\sum_i (x_i - \bar{x})^2}(x - \bar{x})$$

$$\text{or} \quad y - \bar{y} = \frac{\frac{1}{n}\sum_i x_i y_i - \bar{x}\bar{y}}{\frac{1}{n}\sum_i x_i^2 - \bar{x}^2}(x - \bar{x})$$

Notes

1. In the preceding work you will see that only variation in the y values has been considered. The reason for this is that the x values represent a non-random variable. That is why the residuals are vertical and not in any other direction. Thus $y_1, y_2, \ldots$ are values of a random variable Y given by $Y = a + bx + \varepsilon$ where ε is the residual variation, the variation that is not explained by the regression line.

2. The goodness of fit of a regression line may be judged by eye by looking at a scatter diagram. An informal measure which is often used is the coefficient of determination, r^2, which measures the proportion of the total variation in the dependent variable, Y, which is accounted for by the regression line. There is no standard hypothesis test based on the coefficient of determination.

3. This form of the regression line is often called the y *on* x *regression line*. If for some reason you had y as your independent variable, you would use the 'x on y' form obtained by interchanging x and y in the equation.

4. Although the derivation given on pages 139–141 is only true for a random variable on a non-random variable, it happens that, for quite different reasons, the same form of the regression line applies if both variables are random and normally distributed. If this is the case the scatter diagram will usually show an approximately elliptical distribution. Since this is a common situation, this form of the regression line may be used more widely than might at first have seemed to be the case.

EXAMPLE 4.4

A patient is given a drip feed containing a particular chemical and its concentration in his blood is measured, in suitable units, at one hour intervals for the next six hours. The doctors believe the figures to be subject to random errors, arising both from the sampling procedure and the subsequent chemical analysis, but that a linear model is appropriate.

Time, x (hours)	0	1	2	3	4	5	6
Concentration, y	2.4	4.3	5.0	6.9	9.1	11.4	13.5

(i) Find the equation of the regression line of y upon x.
(ii) Estimate the concentration of the chemical in the patient's blood 3 hours 30 minutes after treatment started.

SOLUTION

(i)

x_i	y_i	x_i^2	y_i^2	$x_i y_i$	
0	2.4	0	5.76	0	
1	4.3	1	18.49	4.3	
2	5.0	4	25.00	10.0	
3	6.9	9	47.61	20.7	$n = 7$
4	9.1	16	82.81	36.4	
5	11.4	25	129.96	57.0	
6	13.5	36	182.25	81.0	
Σ 21	52.6	91	491.88	209.4	

$$\bar{x} = \frac{\Sigma x_i}{n} = \frac{21}{7} = 3 \qquad \bar{y} = \frac{\Sigma y_i}{n} = \frac{52.6}{7} = 7.514$$

$$\frac{1}{n}\Sigma x_i^2 - \bar{x}^2 = \frac{91}{7} - 3^2 = 4.0$$

$$\frac{1}{n}\Sigma x_i y_i - \bar{x}\bar{y} = \frac{209.4}{7} - 3 \times 7.514 = 7.372$$

Using $y - \bar{y} = \dfrac{\dfrac{1}{n}\sum_i x_i y_i - \bar{x}\bar{y}}{\dfrac{1}{n}\sum_i x_i^2 - \bar{x}^2}(x - \bar{x})$

the equation of the regression line is given by

$$y - 7.514 = \frac{7.372}{4.0}(x - 3)$$

$$y = 1.843x + 1.986$$

(ii) When $x = 3.5$, $y = 1.843 \times 3.5 + 1.986$
$$= 8.4 \ (1 \ \text{dp})$$

At time 3 hours 30 minutes the concentration is estimated to be 8.4 units.

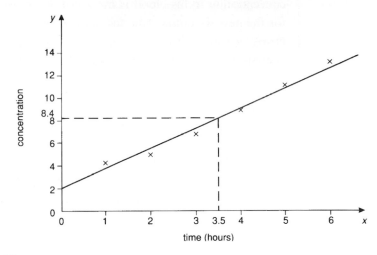

Figure 4.19

Note

The concentration of 8.4 lies between the measured values of 6.9 at time 3 hours and 9.1 at time 4 hours and so seems quite reasonable.

EXERCISE 4E

1 For the following bivariate data obtain the equation of the least squares regression line of y on x. Estimate the value of y when $x = 12$.

x	5	10	15	20	25
y	30	28	27	27	21

2 Calculate the equation of the regression line of y on x for the following distribution and use it to estimate the value of y when $x = 42$.

x	25	30	35	40	45	50
y	78	70	65	58	48	42

3 The 1980 and 2000 catalogue prices in pence of five British postage stamps are as follows:

1980 price, x	10	20	30	40	50
2000 price, y	100	215	280	360	450

(i) Plot these data on a scatter diagram.

(ii) Calculate the equation of the regression line and draw it accurately on your scatter diagram.

(iii) Another stamp was valued at £5 in 1980 and at £62 in 2000. Comment.

4 In an investigation of the genus *Tamarix* (a shrub able to withstand drought), research workers in Tunisia measured the average vigour y (defined as the average width in centimetres of the last two annual rings) and stem density x (defined as the number of stems per m^2) at ten sites with the following results:

x	4	5	6	9	14	15	15	19	21	22
y	0.75	1.20	0.55	0.60	0.65	0.55	0	0.35	0.45	0.40

(i) Draw a scatter diagram for these data.

(ii) Given that $\Sigma x = 130$, $\Sigma x^2 = 2090$, $\Sigma y = 5.5$ and $\Sigma xy = 59.95$, find the regression line of y on x, and plot it on your diagram.

(iii) Use your line to estimate the average vigour for a stem density of 17 stems per m^2.

(iv) Give one reason why it would be invalid to use your regression line to estimate average vigour for stem densities substantially greater than 22 stems per m^2 and explain how your regression line confirms that your reason is correct.

[MEI]

5 Each of a group of 12 apple trees is given the same spraying treatment against codling moth larvae. After the crop is gathered, the apples are examined for grubs with the following results.

Tree number	1	2	3	4	5	6	7	8	9	10	11	12
Size of crop x (hundreds of apples)	8	6	11	22	14	17	18	24	19	23	26	40
Percentage y of apples with grubs	59	58	56	53	50	45	43	42	39	38	30	27

Plot the data on a scatter diagram.

Given that $\Sigma x = 228$, $\Sigma x^2 = 5256$, $\Sigma y = 540$ and $\Sigma xy = 9324$, calculate the line of regression of y on x, and plot the line on your diagram. Estimate to the nearest whole number the expected percentage of apples with grubs for a tree carrying 2000 apples. (You may assume that these procedures are justified because x and y are jointly distributed in a suitable way.)

A second group of 12 apple trees is given a different treatment and the results are given below.

x	15	15	12	26	18	12	8	38	26	19	29	22
y	52	46	38	37	37	37	34	25	22	22	20	14

Plot the data from the second group on the same scatter diagram using a different symbol from that used before. *Without further calculation,* summarize the differences between the results of the two treatments.

[MEI]

6 The speed of a car, v metres per second, at time t seconds after it starts to accelerate is shown in the table below, for $0 \leqslant t \leqslant 10$.

t	0	1	2	3	4	5	6	7	8	9	10
v	0	3.0	6.8	10.2	12.9	16.4	20.0	21.4	23.0	24.6	26.1

$[\Sigma t = 55, \Sigma v = 164.4, \Sigma t^2 = 385, \Sigma v^2 = 3267.98, \Sigma tv = 1117.0.]$

The relationship between t and v is initially modelled by using all the data above and calculating a single regression line.

(i) Plot a scatter diagram of the data, with t on the horizontal axis and v on the vertical axis.

(ii) Using all the data given, calculate the equation of the regression line of v on t. Give numerical coefficients in your answers correct to 3 significant figures.

(iii) Calculate the product moment correlation coefficient for the given data.

(iv) Comment on the validity of modelling the data by a single straight line and on the answer obtained in part (iii).

[Cambridge]

7 The table below shows the names of five toy construction kits which were bought from a catalogue, the numbers of pieces, n, found in each and the corresponding prices paid, £p.

Name	Set 1	Set 3	Set 4	Set 5	Set 6
n	11	21	28	37	75
p	11	26	34	41	88

$[\Sigma n = 172, \Sigma p = 200, \Sigma n^2 = 8340, \Sigma p^2 = 11\,378, \Sigma np = 9736.]$

(i) Plot a scatter diagram of the data, with n on the horizontal axis and p on the vertical axis.

(ii) Calculate the equation of the regression line of p on n, and plot this line on your scatter diagram. Use your equation to estimate the price of Set 2, which is not listed in the catalogue, but is thought to have 15 pieces. Give your answer correct to the nearest pound.

(iii) Calculate the product moment correlation coefficient for the given data, giving your answer correct to 3 decimal places, and interpret the result in terms of your scatter diagram.

[Cambridge]

8 The results of an experiment to determine how the percentage sand content of soil y varies with depth in centimetres below ground level x are given in the following table.

x	0	6	12	18	24	30	36	42	48
y	80.6	63.0	64.3	62.5	57.5	59.2	40.8	46.9	37.6

Calculate

(i) the covariance of x and y

(ii) the product moment correlation coefficient of x and y

(iii) the equation of the line of regression of y on x.

Explain briefly why the product moment correlation coefficient is preferable to the covariance as a measure of the association between x and y.

[MEI]

9 In an investigation into prediction using the stars and planets, a celebrated astrologer Horace Cope predicted the ages at which 13 young people would first marry. The complete data, of predicted and actual ages at first marriage, are now available and are summarised in the following table:

Person	A	B	C	D	E	F	G	H	I	J	K	L	M
Predicted age x (years)	24	30	28	36	20	22	31	28	21	29	40	25	27
Actual age y (years)	23	31	28	35	20	25	45	30	22	27	40	27	26

(i) Draw a scatter diagram of these data.

(ii) Calculate the equation of the regression line of y on x and draw this line on the scatter diagram.

(iii) Comment upon the results obtained, particularly in view of the data for person G. What further action would you suggest?

[AEB]

10 Observations of a cactus graft were made under controlled environmental conditions. The table gives the observed heights y cm of the graft at x weeks after grafting. Also given are the values of $z = \ln y$.

x	1	2	3	4	5	6	8	10
y	2.0	2.4	2.5	5.1	6.7	9.4	18.3	35.1
$z = \ln y$	0.69	0.88	0.92	1.63	1.90	2.24	2.91	3.56

(i) Draw two scatter diagrams, one for y and x, and one for z and x.

(ii) It is desired to estimate the height of the graft seven weeks after grafting. Explain why your scatter diagrams suggest the use of the line of regression of z on x for this purpose, but not the line of regression of y on x.

(iii) Obtain the required estimate given that $\Sigma x = 39$, $\Sigma x^2 = 255$, $\Sigma z = 14.73$, $\Sigma z^2 = 34.5231$, $\Sigma xz = 93.55$.

[MEI]

11 The authorities in a school are concerned to ensure that their students enter appropriate mathematics examinations. As part of a research project into this they wish to set up a performance prediction model. This involves the students taking a standard mid-year test, based on the syllabus and format of the final end-of-year examination.

The school bases its model on the belief that in the final examination students will get the same things right as they did in the mid-year test and in addition a proportion, p, of the things they got wrong.

Consequently a student's final mark, y%, can be predicted on the basis of his or her test mark, x%, by the relationship:

$$y = x + p \times (100 - x)$$

Final mark = Test mark $+ p \times$ (The marks the student did not get on the test)

Investigate this model, using the following bivariate data. Start by finding the y on x regression line and then rearrange it to estimate p.

x	y	x	y	x	y	x	y
40	55	27	44	60	72	46	70
22	40	32	50	50	70	70	85
10	25	26	49	90	95	33	63
46	68	68	76	30	50	40	60
66	75	54	66	64	80	56	57
8	32	68	70	100	100	45	55
48	69	88	92	44	50	78	85
58	66	48	59	58	62	68	80
50	51	82	90	54	60	78	85
80	85	66	76	24	31	89	91

12 A sociologist has this theory: *The proportion of people who get married, 'the marrying kind', is the same across the world. Divorce rates vary widely from one country to another, but many divorcees then remarry. Consequently a country's marriage rate depends on its divorce rate and increases in proportion to it.*

Construct a mathematical model of this theory and test it on the following data for EC countries.

Country	Marriage rate (per 1000 population)	Divorce rate (per 1000 population)
Belgium	6.5	2.0
Denmark	6.1	2.7
Germany	6.5	2.2
Greece	5.4	0.6
Spain	5.5	0.6
France	5.1	1.9
Ireland	5.0	0
Italy	5.4	0.4
Luxemburg	6.1	2.3
Netherlands	6.4	1.9
Portugal	6.9	0.9
UK	6.8	2.9

Source: *The Independent* 23/9/92

Collect a substantial set of raw bivariate data (at least 50 items) relating to a subject which interests you, and present it in a written report. Include in your report:

- suitable graphs
- calculations
- interpretation.

KEY POINTS

1 A scatter diagram is a graph to illustrate bivariate data.

2 Notation for n pairs of observations, (x_i, y_i):

$$S_{xx} = \sum_i (x_i - \bar{x})^2 \qquad S_{yy} = \sum_i (y_i - \bar{y})^2$$

$$S_{xy} = \sum_i (x_i - \bar{x})(y_i - \bar{y})$$

3 Sample covariance $= \dfrac{1}{n} S_{xy} = \dfrac{1}{n} \sum (x_i - \bar{x})(y_i - \bar{y}) = \dfrac{1}{n} \sum x_i y_i - \bar{x}\bar{y}$.

4 The production moment correlation coefficient,

$$r = \frac{S_{xy}}{\sqrt{S_{xx} S_{yy}}}$$

$$\text{or} \quad r = \frac{\sum_i (x_i - \bar{x})(y_i - \bar{y})}{\sqrt{\sum_i (x_i - \bar{x})^2 \sum_i (y_i - \bar{y})^2}}.$$

$$\text{or} \quad r = \frac{\dfrac{1}{n} \sum_i x_i y_i - \bar{x}\bar{y}}{\sqrt{\left(\dfrac{1}{n} \sum_i x_i^2 - \bar{x}^2\right)\left(\dfrac{1}{n} \sum_i y_i^2 - \bar{y}^2\right)}}.$$

5 Spearman's coefficient of rank correlation, $r_s = 1 - \dfrac{6\sum d_i^2}{n(n^2 - 1)}$.

6 The equation of the y on x regression line is

$$y - y = \frac{S_{xy}}{S_{xx}} (x - x)$$

$$\text{or} \quad y - \bar{y} = \frac{\sum_i (x_i - \bar{x})(y_i - \bar{y})}{\sum_i (x_i - \bar{x})^2} (x - \bar{x})$$

$$\text{or} \quad y - \bar{y} = \frac{\dfrac{1}{n} \sum_i x_i y_i - \bar{x}\bar{y}}{\dfrac{1}{n} \sum_i x_i^2 - \bar{x}^2} (x - \bar{x})$$

Appendix

1. Mean and variance of the binomial distribution

> *Note*
>
> This rather lengthy piece of work can be reduced to just a few lines by using probability generating functions, which you meet in *Statistics 5*.

Mean

$$E(X) = \sum_i x_i P(X = x_i) = \sum_r r P(X = r)$$

For the binomial distribution $X \sim B(n, p)$

$$P(X = r) = {}^nC_r p^r (1 - p)^{n-r}$$

$$E(X) = 0 \times {}^nC_0 (1 - p)^n + 1 \times {}^nC_1 p (1 - p)^{n-1}$$

$$+ 2 \times {}^nC_2 p^2 (1 - p)^{n-2} + \cdots + n \, {}^nC_n p^n$$

$$= \sum_{r=0}^{n} r \, {}^nC_r p^r (1 - p)^{n-r}$$

$$= \sum_{r=1}^{n} r \, {}^nC_r p^r (1 - p)^{n-r} \quad \text{(since the first term is zero)}$$

$$= \sum_{r=1}^{n} r \frac{n!}{r!(n - r)!} p^r (1 - p)^{n-r}$$

$$= \sum_{r=1}^{n} \frac{n!}{(r - 1)!(n - r)!} p^r (1 - p)^{n-r}$$

$$= np \sum_{r=1}^{n} \frac{(n - 1)!}{(r - 1)!(n - r)!} p^{r-1} (1 - p)^{n-r}$$

$$= np \sum_{r=1}^{n} {}^{n-1}C_{r-1} p^{r-1} (1 - p)^{n-r}$$

But $\displaystyle\sum_{r=1}^{n} {}^{n-1}C_{r-1} p^{r-1} (1 - p)^{n-r} = [p + (1 - p)]^{n-1} = 1$

$$\therefore E(X) = np$$

Variance

$$\text{Var}(X) = \sum_i x_i^2 P(X = x_i) - [E(X)]^2$$

$$= \sum_{r=0}^{n} r^2 \, {}^nC_r p^r (1 - p)^{n-r} - n^2 p^2$$

$$= \sum_{r=1}^{n} r^2 \frac{n!}{r!(n - r)!} \, p^r (1 - p)^{n-r} - n^2 p^2 \quad \text{(since the first term is zero)}$$

$$= np \sum_{r=1}^{n} r \frac{(n - 1)!}{(r - 1)!(n - r)!} \, p^{r-1}(1 - p)^{n-r} - n^2 p^2$$

$$= np \left[\sum_{r=1}^{n} \left\{ \frac{(r - 1)(n - 1)!}{(r - 1)!(n - r)!} + \frac{(n - 1)!}{(r - 1)!(n - r)!} \right\} p^{r-1}(1 - p)^{n-r} \right] - n^2 p^2$$

$$= np \left[\sum_{r=2}^{n} (r - 1) \frac{(n - 1)!}{(r - 1)!(n - r)!} \, p^{r-1}(1 - p)^{n-r} \right.$$

$$\left. + \sum_{r=1}^{n} \frac{(n - 1)!}{(r - 1)!(n - r)!} \, p^{r-1}(1 - p)^{n-r} \right] - n^2 p^2$$

$$= np \left[(n - 1)p \sum_{r=2}^{n} \frac{(n - 2)!}{(r - 2)!(n - r)!} \, p^{r-2}(1 - p)^{n-r} \right.$$

$$\left. + \{ p + (1 - p) \}^{n-1} \right] - n^2 p^2$$

$$= np[(n - 1)p\{[p + (1 - p)]^{n-2}\} + 1] - n^2 p^2$$

$$= np[(n - 1)p + 1] - n^2 p^2$$

$$= n^2 p^2 - np^2 + np - n^2 p^2$$

$$= np(1 - p)$$

$$= npq$$

2. Mean and variance of the Poisson distribution

Mean

$$E(X) = \sum_i x_i P(X = x_i) = \sum_{r=0}^{\infty} r P(X = r)$$

$$= 0 + 1 \times P(X = 1) + 2 \times P(X = 2) + 3 \times P(X = 3) + \cdots$$

$$= \frac{1.\lambda e^{-\lambda}}{1!} + \frac{2.\lambda^2 e^{-\lambda}}{2!} + \frac{3.\lambda^3 e^{-\lambda}}{3!} + \cdots$$

$$= \lambda e^{-\lambda} \left(1 + \lambda + \frac{\lambda^2}{2!} + \frac{\lambda^3}{3!} + \cdots \right)$$

The series in the brackets $= e^{\lambda}$

$$\therefore E(X) = \lambda e^{-\lambda} e^{\lambda}$$

$$= \lambda$$

Variance

$$\text{Var}(X) = E(X^2) - [E(X)]^2$$

$$E(X^2) = \sum_{r=0}^{\infty} r^2 P(X = r)$$

Putting
$$r^2 = r(r - 1) + r$$

gives
$$E(X^2) = \sum_{r=0}^{\infty} \{r(r - 1) + r\} P(X = r)$$

$$= \sum_{r=0}^{\infty} r(r - 1) P(X = r) + \sum_{r=0}^{\infty} r P(X = r)$$

Now
$$\sum_{r=0}^{\infty} r P(X = r) = \lambda$$

$$\therefore E(X^2) = \sum_{r=0}^{\infty} r(r - 1) P(X = r) + \lambda$$

Now the first term, $\displaystyle\sum_{r=0}^{\infty} r(r - 1) P(X = r)$

$$= 0 + 0 + 2 \times 1 \times P(X = 2) + 3 \times 2 \times P(X = 3)$$

$$+ 4 \times 3 \times P(X = 4) + \cdots$$

$$= \frac{2 \times 1 \times \lambda^2 e^{-\lambda}}{2!} + \frac{3 \times 2 \times \lambda^3 e^{-\lambda}}{3!}$$

$$+ \frac{4 \times 3 \times \lambda^4 e^{-\lambda}}{4!} + \cdots$$

Cancelling
$$= \lambda^2 e^{-\lambda} + \lambda^3 e^{-\lambda} + \frac{\lambda^4 e^{-\lambda}}{2!} + \frac{\lambda^5 e^{-\lambda}}{3!} + \cdots$$

$$= \lambda^2 e^{-\lambda} \left(1 + \lambda + \frac{\lambda^2}{2!} + \frac{\lambda^3}{3!} + \cdots \right)$$

Once again the term in brackets $= e^{\lambda}$

$$= \lambda^2 e^{-\lambda} e^{\lambda}$$

$$= \lambda^2$$

This now gives:

$$E(X^2) = \lambda^2 + \lambda$$

So that:

$$\text{Var}(X) = \lambda^2 + \lambda - \lambda^2$$

$$= \lambda$$

Thus showing that for the Poisson distribution the mean and variance are both λ.

3. The sum of two independent Poisson distributions

Suppose that $X \sim$ Poisson (λ) and $Y \sim$ Poisson (μ) and that X and Y are independent, and that T is the random variable given by

$$T = X + Y$$

$$P(T = t) = P(X = t) \times P(Y = 0) + P(X = t - 1) \times P(Y = 1) + P(X = t - 2) \\ \times P(Y = 2) + \cdots + P(X = 0) \times P(Y = t)$$

X and Y must be independent so that the probabilities can be multiplied.

$$= \frac{\lambda^t e^{-\lambda} e^{-\mu}}{t!} + \frac{\lambda^{t-1} e^{-\lambda} \mu e^{-\mu}}{(t-1)!1!} + \frac{\lambda^{t-2} e^{-\lambda} \mu^2 e^{-\mu}}{(t-2)!2!} + \cdots + \frac{e^{-\lambda} \mu^t e^{-\mu}}{t!}$$

$$= e^{-(\lambda+\mu)} \left\{ \frac{\lambda^t}{t!} + \frac{\lambda^{t-1} \mu}{(t-1)!1!} + \frac{\lambda^{t-2} \mu^2}{(t-1)!2!} + \cdots + \frac{\mu^t}{t!} \right\}$$

$$= \frac{e^{-(\lambda+\mu)}}{t!} \left\{ \lambda^t + \frac{t! \lambda^{t-1} \mu}{(t-1)!1!} + \frac{t! \lambda^{t-2} \mu^2}{(t-1)!2!} + \cdots + \mu^t \right\}$$

Now the term in brackets is just the binomial expansion of $(\lambda + \mu)^t$

$$P(T = t) = \frac{e^{-(\lambda+\mu)} (\lambda + \mu)^t}{t!}$$

which is the Poisson probability with parameter $\lambda + \mu$.

So the sum of the two independent Poisson distributions with parameters λ and μ is itself a Poisson distribution with parameter $\lambda + \mu$.

Similarly, the sum of n independent Poisson distributions with parameters $\lambda_1, \lambda_2, \ldots, \lambda_n$ is a Poisson distribution with parameter $(\lambda_1 + \lambda_2 + \cdots + \lambda_n)$.

4. Equivalence of Spearman's rank correlation coefficient and Pearson's product moment correlation coefficient

Let the ranks assigned to n items by two judges be $x_1, x_2, \ldots, x_n$ and $y_1, y_2, \ldots, y_n$

	x_1	x_2	$\cdots$	x_i	$\cdots$	x_n
	y_1	y_2	$\cdots$	y_i	$\cdots$	x_n
$\|d_i\| =$	$\|x_1 - y_1\|$	$\|x_2 - y_2\|$	$\cdots$	$\|x_i - y_i\|$	$\cdots$	$\|x_n - y_n\|$

$$\sum_{i=1}^{n} d_i^2 = \sum_{i=1}^{n} (x_i - y_i)^2 = \sum_i x_i^2 + \sum_i y_i^2 - 2 \sum_i x_i y_i$$

Since the values of $x_1, \ldots, x_n$ and $y_1, \ldots, y_n$ are both $1, 2, 3, \ldots, n$ in some order

$$\sum_{i=1}^{n} x_i = \sum_{i=1}^{n} y_i = \frac{n(n+1)}{2}$$

$$\bar{x} = \bar{y} = \frac{n+1}{2}$$

$$\sum_{i=1}^{n} x_i^2 = \sum_{i=1}^{n} y_i^2 = \frac{n(n+1)(2n+1)}{6}$$

$$\frac{1}{n} \sum_{i=1}^{n} x_i^2 - \bar{x}^2 = \frac{n(n+1)(2n+1)}{6n} - \left(\frac{n+1}{2}\right)^2$$

$$= \frac{(n+1)(n-1)}{12} = \frac{n^2-1}{12} = \frac{1}{n} \sum_{i=1}^{n} y_i^2 - \bar{y}^2$$

$$r_s = 1 - \frac{6\sum d_i^2}{n(n^2-1)}$$

$$= 1 - \frac{6[\sum x_i^2 + \sum y_i^2 - 2\sum x_i y_i]}{n(n^2-1)}$$

$$= \frac{1}{n(n^2-1)} \times \left\{ n(n^2-1) \right.$$

$$\left. - 6 \left[\frac{n(n+1)(2n+1)}{6} + \frac{n(n+1)(2n+1)}{6} - 2\sum x_i y_i \right] \right\}$$

$$= \frac{12\sum x_i y_i - 3n(n+1)^2}{n(n^2-1)}$$

$$= \frac{12 \left[\frac{\sum x_i y_i}{n} - \left(\frac{n+1}{2}\right)^2 \right]}{(n^2-1)}$$

$$= \frac{\sum \frac{x_i y_i}{n} - \bar{x}\bar{y}}{\sqrt{\left(\frac{n^2-1}{12}\right)\left(\frac{n^2-1}{12}\right)}}$$

$$= r \qquad \text{as required}$$

5. The least squares regression line

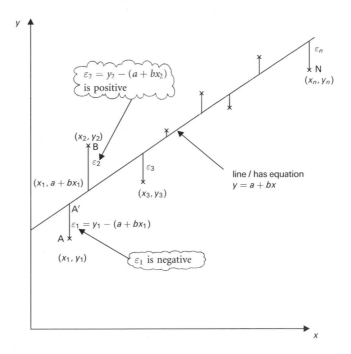

Figure A

The least squares regression line for the set of bivariate data $(x_1, y_1), (x_2, y_2), \ldots,$ (x_n, y_n) is of the form $y = a + bx$ with the values of a and b giving the minimum sum of the squares of the residuals, $\varepsilon_1, \varepsilon_2, \ldots, \varepsilon_n$.

Since $\varepsilon_r = y_r - a + bx_r$, the values of a and b which minimise

$$T = \sum_{r=1}^{n} (y_r - a + bx_r)^2 \text{ must be found.}$$

This is the same as $T = \sum_{r=1}^{n} (a + bx_r - y_r)^2$. The proof which follows uses this form (which, with a at the front, is perhaps slightly easier to follow).

$$T = \sum_{r=1}^{n} (a + bx_r - y_r)^2$$

$$= \sum_{r=1}^{n} [a + (bx_r - y_r)]^2$$

$$= \sum_{r=1}^{n} [a^2 + 2a(bx_r - y_r) + (bx_r - y_r)^2]$$

$$= na^2 + 2a \sum_{r=1}^{n} (bx_r - y_r) + \sum_{r=1}^{n} (bx_r - y_r)^2$$

$$= n\left[a^2 + \frac{2a}{n} \sum_{r=1}^{n} (bx_r - y_r) + \frac{1}{n} \sum_{r=1}^{n} (bx_r - y_r)^2\right]$$

You can treat the right-hand side of the above expression as a quadratic in *a*. Completing the square on it

$$T = n\left\{ a^2 + \frac{2a}{n}\sum_{r=1}^{n}(bx_r - y_r) + \frac{1}{n^2}\left[\sum_{r=1}^{n}(bx_r - y_r)\right]^2 \right.$$

$$\left. + \frac{1}{n}\sum_{r=1}^{n}(bx_r - y_r)^2 - \frac{1}{n^2}\left[\sum_{r=1}^{n}(bx_r - y_r)\right]^2 \right\}$$

$$= n\left[a + \frac{1}{n}\sum_{r=1}^{n}(bx_r - y_r)\right]^2 + \sum_{r=1}^{n}(bx_r - y_r)^2 - \frac{1}{n}\left[\sum_{r=1}^{n}(bx_r - y_r)\right]^2$$

$$= n\left[a - \frac{1}{n}\sum_{r=1}^{n}(y_r - bx_r)\right]^2 + \sum_{r=1}^{n}(bx_r - y_r)^2 - \frac{1}{n}\left[\sum_{r=1}^{n}(bx_r - y_r)\right]^2$$

The last two terms do not involve *a* and so if we wish to choose *a* so as to minimise *T* it seems sensible to let

$$a = \frac{1}{n}\sum_{r=1}^{n}(y_r - bx_r),$$

for then the first term on the right-hand side becomes zero and this term was of course a non-negative one.

This means

$$a = \frac{1}{n}\sum_{r=1}^{n}y_r - \frac{b}{n}\sum_{r=1}^{n}x_r.$$

$$\therefore a = \bar{y} - b\bar{x} \qquad \qquad ①$$

Since the equation of our required regression line is $y = ax + b$, equation ① just means that we wish our regression line to pass through $(\bar{x}, \bar{y})$. This is clearly sensible since $(\bar{x}, \bar{y})$ is the mean point and so approximately the centre of the scatter diagram.

$$T = \sum_{r=1}^{n}(a + bx_r - y_r)^2$$

Substitute now for *a* from equation ①,

$$T = \sum_{r=1}^{n}(\bar{y} - b\bar{x} + bx_r - y_r)^2$$

By reversing all signs

$$T = \sum_{r=1}^{n} [(y_r - \bar{y}) - b(x_r - \bar{x})]^2$$

$$= \sum_{r=1}^{n} (y_r - \bar{y})^2 - 2b \sum_{r=1}^{n} (x_r - \bar{x})(y_r - \bar{y}) + b^2 \sum_{r=1}^{n} (x_r - \bar{x})^2$$

$$= S_{yy} - 2bS_{xy} + b^2 S_{xx}$$

$$= S_{xx} \left[b^2 - 2b \frac{S_{xy}}{S_{xx}} + \frac{S_{yy}}{S_{xx}} \right]$$

Now you can treat the square bracket as a quadratic in b. Again, complete the square.

$$T = S_{xx} \left[b^2 - 2b \frac{S_{xy}}{S_{xx}} + \left(\frac{S_{xy}}{S_{xx}} \right)^2 + \frac{S_{yy}}{S_{xx}} - \left(\frac{S_{xy}}{S_{xx}} \right)^2 \right]$$

$$= S_{xx} \left[\left(b - \frac{S_{xy}}{S_{xx}} \right)^2 + \frac{S_{yy}}{S_{xx}} - \left(\frac{S_{xy}}{S_{xx}} \right)^2 \right]$$

The last two terms on the right-hand side are constants for the distribution and so to minimise T you set the first term in the square bracket on the right-hand side equal to zero, i.e.

$$b = \frac{S_{xy}}{S_{xx}}$$

The regression line to minimise T is now

$$y = a + bx$$

$$= (\bar{y} - b\bar{x}) + bx$$

$$\therefore y - \bar{y} = b(x - \bar{x})$$

$$\therefore y - \bar{y} = \frac{S_{xy}}{S_{xx}} (x - \bar{x})$$

and this equation is called the least squares regression line of y on x.

Answers

Chapter 1

❓ **(Page 1)**
See text that follows.

❓ **(Page 1)**
See text that follows.

Exercise 1A (Page 8)

1 $k = 1/10$

2 Number of turns needed to obtain a '6' on a die.

3 (i) $a = 0.4$ (ii) 0.3

4

X	0	1	2	3	4	5
Probability	$\frac{1}{32}$	$\frac{5}{32}$	$\frac{10}{32}$	$\frac{10}{32}$	$\frac{5}{32}$	$\frac{1}{32}$

5 $c = \frac{1}{28}, \frac{3}{28}$

6

X	2	3	4	5	6	7	8	9	10	11	12
Probability	$\frac{1}{36}$	$\frac{2}{36}$	$\frac{3}{36}$	$\frac{4}{36}$	$\frac{5}{36}$	$\frac{6}{36}$	$\frac{5}{36}$	$\frac{4}{36}$	$\frac{3}{36}$	$\frac{2}{36}$	$\frac{1}{36}$

7 (i)

Y	0	1	2	3	4	5
Probability	$\frac{6}{36}$	$\frac{10}{36}$	$\frac{8}{36}$	$\frac{6}{36}$	$\frac{4}{36}$	$\frac{2}{36}$

(ii) $\frac{2}{3}$

8 (i)

Z	0	1	2	3	4
Probability	$\frac{1}{16}$	$\frac{4}{16}$	$\frac{6}{16}$	$\frac{4}{16}$	$\frac{1}{16}$

(ii) $\frac{5}{16}$

9

X	0	1	2	3
Probability	0.1667	0.5	0.3	0.0333

10

Number of men	0	1	2	3
Probability	0.122	0.441	0.367	0.070

11 (i)

X	1	2	3	4	6	8	9	12	16
Probability	$\frac{1}{16}$	$\frac{2}{16}$	$\frac{2}{16}$	$\frac{3}{16}$	$\frac{2}{16}$	$\frac{2}{16}$	$\frac{1}{16}$	$\frac{2}{16}$	$\frac{1}{16}$

(ii) 0.25

12

Number of red cards	0	1	2	3	4
Probability	0.055	0.25	0.39	0.25	0.055

13 (i) $k = 0.08$

X	0	1	2	3	4
Probability	0.2	0.24	0.32	0.24	0

(ii)

Number of chicks surviving	0	1	2	3
Probability	0.35104	0.44928	0.18432	0.01536

14 (i) $a = 0.42$ (ii) $k = \frac{1}{35}$ (iii) Algebraic model is not particularly accurate. No.

15 (i) $\frac{1}{1296}$ (ii) $\frac{1}{81}$ (iii) $\frac{65}{1296}$ (iv) $\frac{671}{1296}$, 6 is the most likely value because $P(X = 6) > \frac{1}{2}$.

❓ **(Page 15)**
You will learn on page 20 that this difference is the variance of random variable X.

Exercise 1B (Page 16)

1 1.5

2 2.7

3 $P(X = 4) = 0.8$, $P(X = 5) = 0.2$

4

Y	50	100
Probability	0.4	0.6

5 3.22

6 (i) 0.65   (ii) 0.5   (iii) 0.4

7 (i) loss of 2.5p   (ii) loss of 7.5p   (iii) loss of £2.50

8 (i) £168   (ii) £2016

9 (i) (a) 2.79   (b) 8.97   (c) 20.94

10 (i) $w = 21$ (ii) $w = 27.67$

11 (i)

X	2	3	4	5	6	7	8	9	10	11	12
$P(X)$	$\frac{1}{36}$	$\frac{2}{36}$	$\frac{3}{36}$	$\frac{4}{36}$	$\frac{5}{36}$	$\frac{6}{36}$	$\frac{5}{36}$	$\frac{4}{36}$	$\frac{3}{36}$	$\frac{2}{36}$	$\frac{1}{36}$

(ii)

Y	2	4	6	8	10	12
$P(Y)$	$\frac{1}{6}$	$\frac{1}{6}$	$\frac{1}{6}$	$\frac{1}{6}$	$\frac{1}{6}$	$\frac{1}{6}$

(iii) (a) $E(X) = E(Y) = 7$
(b) Range of X = range of $Y = 10$
(c) Mode of $X = 7$, Y has no mode since all the outcomes are equally likely.

12 (i) 3 coins   **(ii)** £1.30

13

p_1	p_2	p_3	p_4	p_5	p_6
$\frac{1}{36}$	$\frac{3}{36}$	$\frac{5}{36}$	$\frac{7}{36}$	$\frac{9}{36}$	$\frac{11}{36}$

(i) 0.0405   **(ii)** 0.1118

14 $\frac{4}{35}, \frac{18}{35}, \frac{12}{35}, \frac{1}{35}, \frac{£(54+r)}{35}$, $r = 50$

15 (i) (a) $\frac{32}{80}$ **(b)** $\frac{36}{48}$ **(c)** $\frac{3}{12}$ **(ii)** $\frac{9}{80}$ **(iii)** $\frac{36}{71}$ **(iv)** £563.38
(v) One week, £625; two week, £494.12. Two weeks.

Exercise 1C (Page 24)

1 (i) (a) $E(X) = 3.1$ **(b)** $Var(X) = 1.29$

2 (i) (a) $E(X) = 0.7$ **(b)** $Var(X) = 0.61$

4 (i) $E(2X) = 6$ **(ii)** $Var(3X) = 6.75$

5 (i) $k = 0.1$ **(ii)** 1.25 eggs **(iii)** 0.942

6 (i) $E(Y) = 1.2857$ **(ii)** $Var(Y) = 0.49$

7 (i) 10.9, 3.09 **(ii)** 18.4, 111.24

8 (i) 2 **(ii)** 1 **(iii)** 9

9 £5.47, 0.086

10

S	1	2	3	6	10
Probability	$\frac{1}{6}$	$\frac{1}{3}$	$\frac{1}{6}$	$\frac{1}{6}$	$\frac{1}{6}$

4, $9\frac{2}{3}$

11 (i) 2.3, 0.41

(ii)

$S_1 + S_2$	2	3	4
Probability	0.51	0.45	0.04

2.53

12 (ii)

X	0	1	2	3	4	6
Probability	$\frac{1}{4}$	$\frac{1}{3}$	$\frac{1}{9}$	$\frac{1}{6}$	$\frac{1}{9}$	$\frac{1}{36}$

(iii) $1\frac{2}{3}, 2\frac{5}{18}$

13

X	6	7	8	9	10
Probability	$\frac{6}{72}$	$\frac{24}{72}$	$\frac{24}{72}$	$\frac{16}{72}$	$\frac{2}{72}$

0.975, 0.6397

14 2.449, 2.574

15

N	2	3	4	5	6	7	8
Probability	$\frac{1}{16}$	$\frac{2}{16}$	$\frac{3}{16}$	$\frac{4}{16}$	$\frac{3}{16}$	$\frac{2}{16}$	$\frac{1}{16}$

5, 2.5

16 (i)

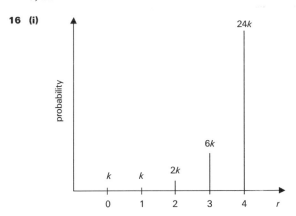

(ii) 3.5, 0.0897 **(iii)** 0.5346 **(iv)** 0.932

17 (i)

X	0	1	2	3	4
Probability	0.003	0.1	0.133	0.167	0.167

X	5	6	7	8
Probability	0.167	0.122	0.089	0.022

(ii) 3.9, 3.93 **(iii)** $k = \frac{1}{84}$ **(iv)** 4, 3 **(v)** Yes

18 (i)

Drinks per customer	1	2	3	4	5	6
Likelihood to the nearest per cent	29%	24%	19%	14%	10%	5%

(iii) 2.67, 2.22
(iv) If the expectation value is large, sales of wine will be good. If the expectation value is small, sales will be poor.
If the value of the variance is large there will be a lot of spread in demand so that total sales on any one day will be somewhat unpredictable. If the value of the variance is small then sales will be fairly consistent.
(v) a: 1.62, 0.52; b: 1.336, 0.70
(vi) The variance of both models is significantly less than with the original model. This suggests that demand for drinks will be fairly consistent. Serving double measures will reduce pressure on the bar staff. Model (a) will lead to a slight increase in sales over the original model. Model (b) will result in approximately the same consumption of wine but nearly 15% of the original customers will no longer frequent the wine bar.

19 (i)

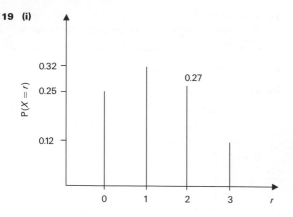

(ii) 1.34, 1.044

(iii) **(a)** 0.0625 **(b)** 0.1753. Assume the number of goals scored is independent of previous results.

(iv) Injuries to players, new signings, promotion to a new league, etc. Model does not allow more than 3 goals to be scored.

20 (i)

r	1	2	3	4	5
$P(X = r)$	$30k$	$20k$	$12k$	$6k$	$2k$

(ii)

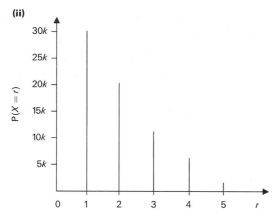

(iii) 2, 1.095

(iv) £1.70

Exercise 1D (Page 32)

1 (i)

r	0	1	2	3	4
$P(R = r)$	$\frac{1}{16}$	$\frac{4}{16}$	$\frac{6}{16}$	$\frac{4}{16}$	$\frac{1}{16}$

(ii) 2, 1

2 (i)

x	0	1	2	3
$P(X = x)$	0.5787	0.3472	0.0694	0.00463

(ii) 0.5, 0.417

3 (i) 5, 2.5 **(ii)** 5.15, 1.8275 **(iii)** Yes

4 (i)

Y	5	10	15	20	25	30
Probability	$\frac{1}{6}$	$\frac{1}{6}$	$\frac{1}{6}$	$\frac{1}{6}$	$\frac{1}{6}$	$\frac{1}{6}$

(ii) £17.50, £8.54

(iii) Perhaps somebody has donated an additional £1000!

5 (i) 2 **(ii)** 3

(iii) First couple: $\frac{1}{8}$; second couple: $\frac{1}{4}$

6 (i) Geometric distribution with $p = 0.4$

(ii) 2.5, 3.75

7 (i)

X	1	2	3	4
Probability	$\frac{1}{4}$	$\frac{1}{4}$	$\frac{1}{4}$	$\frac{1}{4}$

2.5, 1.25

(ii)

Y	2	3	4	5	6	7	8
Probability	$\frac{1}{16}$	$\frac{2}{16}$	$\frac{3}{16}$	$\frac{4}{16}$	$\frac{3}{16}$	$\frac{2}{16}$	$\frac{1}{16}$

5, 2.5

(iii) 3 packets **(iv)** 0.859

8

Number of heads, X	0	1	2	3
Probability	q^3	$3q^2p$	$3qp^2$	p^3

1.66, $p = 0.553$; 0.982

9 (i)

r	0	1	2	3
$P(X \leqslant r)$	0.1	0.5	0.8	1.0

(ii) If the largest individual score, L, is at most 2 then each of the three individual scores, X, must be at most 2. Thus $P(L \leqslant 2) = [P(X \leqslant 2)]^3 = (0.8)^3 = 0.512$.
0.125; 0.001, 1

(iii)

r	0	1	2	3
$P(L = r)$	0.001	0.124	0.387	0.488

(iv) 2.362, 0.485

10 (i) 0.3, 0.147, 0.343

(ii)

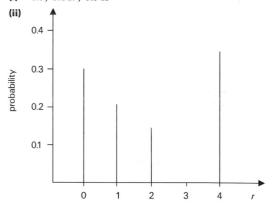

(iii) 1.876, 1.66 **(iv)** 0.363

Chapter 2

Exercise 2A (Page 46)

1 **(i)** 0.271 **(ii)** 0.090

2 **(i)** 0.082 **(ii)** 0.214 **(iii)** 0.067

3 **(i)** 0.050 **(ii)** 0.149 **(iii)** 0.224 **(iv)** 0.423 **(v)** 0.577

4 **(i)** 0.359 **(ii)** 0.641

5 0.054

6 X may be modelled by a Poisson distribution when cars arrive singly and independently and at a known overall average rate.
0.442

7 0.79
It is assumed that calls arrive singly and independently and with a known overall average rate of 4.2 calls per night.

8 **(i)** 25 **(ii)** 75 **(iii)** 112 **(iv)** 278
Assume that mistakes occur randomly, singly, independently and at a constant mean rate.

9 **(i)** 42 **(ii)** 2, 2.381
(iii) 0.135, 0.271, 0.271, 0.180, 0.090, 0.036, 0.012
(iv) 5.7, 11.4, 11.4, 7.6, 3.8, 1.5, 0.5
(v) Yes, because there is a good fit between the actual data and the predictions made in part (iv).

10 **(i)** 0.111, 0.244, 0.268, 0.377 **(ii)** 3

11 **(i)** 0.135 **(ii)** 0.271 **(iii)** 36 tubs **(iv)** 2.6 complaints

12 **(i)** 0.738 **(ii)** 239, 177, 65, 16, 0, 0
(iii) 239.0, 176.4, 65.1, 16.0, 3.0, 0.4
(iv) Yes, there seems to be reasonable agreement between the actual data and the Poisson predictions.

13 **(i)** 0.3328 **(ii)** 0.0016

14 600 m;
Poisson distribution with parameter 2.5;
0.082, 0.109; 0.779, 0.207

15 Because we assume injuries occur singly, independently and randomly.
0.5, 0.48; because the mean is approximately equal to the variance.
31.5, 15.8, 3.9, 0.7, 0.1

16 **(i)** 30 **(ii)** 27.5 **(iii)** 461

17 7.03, 6.84;
The mean is approximately equal to the variance.
8.17

18 **(i)** 0.165 **(ii)** 5 **(iii)** 0.027 **(iv)** 5

19 **(i)** The mean is much greater than the variance therefore X does not have a Poisson distribution.
(ii) Yes because now the values of the mean and variance are similar.
(iii) 0.012

20 **(i)** X can be modelled by a Poisson distribution because the passage of cars happens independently, at random and at a constant mean rate. Also the values of the mean and variance are similar.
(ii) 0.224 **(iii)** 0.586

❓ (Page 53)

1 It is not necessarily so that a car or lorry passing along the road is a random event. Regular users will change both Poisson parameters which in turn will affect the solution to the problem.

2 No. Traffic tends to travel in a line at the same speed on some roads.

3 It could be that their numbers are negligible or it might be assumed they do not damage the cattle grid.

Exercise 2B (Page 53)

1 **(i)** 0.362 **(ii)** 0.544 **(iii)** 0.214 **(iv)** 0.558

2 **(i)** 0.082 **(ii)** 0.891 **(iii)** 0.287

3 **(i)** 0.161 **(ii)** 0.554 **(iii)** 10 **(iv)** 0.016 **(v)** 0.119

4 **(i)** 0.102 **(ii)** 0.285 **(iii)** 0.422

5 **(i)** 0.175 **(ii)** 0.973 **(iii)** 0.031; 0.125; 0.249

6 **(i)** 0.175 **(ii)** 0.560 **(iii)** 0.1251 **(iv)** 0.5421
(v) 0.0308; 10

7 Some bottles will contain two or more hard particles. This will decrease the percentage of bottles that have to be discarded.
13.9%
Assume the hard particles occur singly, independently and randomly.

8 $X \sim P(1)$; 0.014; 0.205

9 **(i)** 0.135 **(ii)** 0.9473 **(iii)** 0.0527 **(iv)** 13 **(v)** 0.1242

10 **(i) (a)** 0.082 **(b)** 0.456 **(ii)** 0.309 **(iii)** 1 east bound and 2 west-bound

11 3.87, 3.53, taking $f_9 = 8$

Poisson distribution with parameter 3.87 because radioactive atoms decay randomly and independently and at a constant mean rate if the half-life is long compared with the duration of the experiment.

10.8, 41.9, 81.1, 104.6, 101.2, 78.3, 50.5, 27.9. 13.5, 9.2

There is quite good agreement between the two sets of figures.

75.8

12 (i) 0.27 **(ii)** 0.3504 **(iii)** 0.182

(iv) 0.1236; demand can be met provided not more than three cars are requested on any one day.

(v) £32.42

13 (i) It is assumed that incidents of criminal damage occur singly, independently and at a constant mean rate.

(ii) (a) 0.271 **(b)** 0.018 **(c)** 0.184

14 (i) 0.47 **(ii)** 0.04

The binomial model assumes that the probability of dialling a wrong number is the same for whatever phone call is made. This is questionable since one might expect a greater likelihood of an error when dialling long distance or unfamiliar numbers than when dialling local or frequently used numbers. It seems sensible to use the Poisson approximation since n is large and p is small and the corresponding Poisson parameter, 0.75, is simple to use.

15 (i) (a) The distribution of X is approximately binomial, n is large, p is small and $\lambda = np = 20$.

(b) 0.86

(ii) 230

Chapter 3

Exercise 3A (Page 70)

1 (i) 0.8413 **(ii)** 0.0228 **(iii)** 0.1359

2 (i) 0.0668 **(ii)** 0.6915 **(iii)** 0.2417

3 (i) 0.0668 **(ii)** 0.1587 **(iii)** 0.7745

4 (i) 0.0038 **(ii)** 0.5 **(iii)** 0.495

5 (i) 31, 1.97 **(ii)** 2.1, 13.4, 34.5, 34.5, 13.4, 2.1

(iii) More data would need to be taken to say reliably that the weights are normally distributed.

6 0.525

7 (i) 5.48% **(ii) (a)** 25 425 km **(b)** 1843 km

8 (i) 78.65% **(ii)** 5.254, 0.054

9 (i) 20.3% **(ii)** 81.0 g

10 (i) 0.0765 **(ii)** 0.2216 **(iii)** 0.4315

11 (i) 0.077 **(ii)** 0.847 **(iii)** 0.674 **(iv)** 1.313 m

12 (i) 0.0900 **(ii)** 0.5392 **(iii)** 0.3467 **(iv)** 8.28 and 30 s

13 0.0401 **(i)** 0.4593 **(ii)** 0.003

14 (i) (a) 0.675 **(b)** 0.325 **(ii)** 0.481 **(iii)** 31 days

15 20.05, 0.0248, 0.7794, 22.6%

16 (i) 0.4875 **(ii)** 281, 5.00

17 0.0312 **(i)** 0.927 **(ii)** 0.08

18 (i) (a) B **(b)** A **(ii)** A(£15.16), B(£10.96)

19 11.09%, 0.0123, 0.0243

20 (i) 0.0668 **(ii)** 0.7745 **(iii)** 0.819

(iv) $\mu > 266.45$ so say 267 ml

❷ (Page 78)

One possibility is that some people, knowing their votes should be secret, resented being asked who they had supported and so deliberately gave wrong answers. Another is that the exit poll was taken at a time of day when those voting were unrepresentative of the electorate as a whole.

Exercise 3B (Page 80)

1 99.4% of the population have IQs less than 2.5 standard deviations above the mean.

0.165

2 (i) 106 **(ii)** 75 and 125 **(iii)** 39.5

3 (i) 0.282 **(ii)** 0.526 **(iii) (a)** 25 **(b)** 4.33 **(iv)** 0.102

4 0.246, 0.0796, 0.0179

The normal distribution is used for continuous data; the binomial distribution is used for discrete data. If a normal approximation to the binomial distribution is used then a continuity correction must be made. Without this the result would not be accurate.

5 n must be large and p must not be too close to 0 or 1. These conditions ensure that the distribution is reasonably symmetrical so that its probability profile resembles a normal distribution.

(i) 0.1853 **(ii)** 0.1838 **(iii)** 0.81%

6 **(i)** **(a)** 0.315 **(b)** 0.307; assuming the answer to part
(i)(a) is correct, there is a 7.6% error; worse
(ii) 0.5245

7 $\frac{1}{3}$; 6.667; 4.444; 13

8 **(i)** 0.590 **(ii)** 0.081
It is assumed that the defective syringes are mixed
randomly with the functional ones. Also, as the total
number of syringes in the box is very large, removing
one syringe leaves the probability distribution of
defective and functional syringes unchanged.

It may be convenient to use the normal
approximation as this simplifies the probability
calculation. So long as a continuity correction is
made the result ought to be very close to the binomial
model result.
$P(X \geqslant 15) = 0.067$

9 **(i)** 0.2305 **(ii)** $\lambda = 9.2$; n is large and p is small.
(iii) 0.3180 **(iv)** 15 **(v)** It seems inconsistent with
the assumptions made earlier because
$P(X \geqslant 20) = 0.0014$. The assumption of
independence needs questioning.

10 $^{2000}C_N \left(\dfrac{1}{30}\right)^N \left(\dfrac{29}{30}\right)^{2000-N}$

86; more (96)

11 0.180; 60, 7.75, 0.9124

12 Mean $= \displaystyle\sum_{r=0}^{\infty} rP(X = r) = \lambda$, called the Poisson
parameter.

Variance $= E(X^2) - \lambda^2 = \lambda$
The Poisson parameter should be greater than 10 so
that the probability profile is approximately bell-
shaped.
(i) 0.0222 **(ii)** 0.9778

13 **(i)** X has a binomial distribution with $p = 0.8$
i.e. $X \sim B(n, 0.8)$; 0.9389
(ii) $X \sim N(80, 16)$; 0.0845
(iii) 71

14 **(i)** 2.5; assume that service calls occur singly,
independently and randomly.
(ii) 0.918, 0.358 **(iii)** 0.158

Chapter 4

❓ **(Page 89)**
See text that follows

Exercise 4A (Page 95)

1
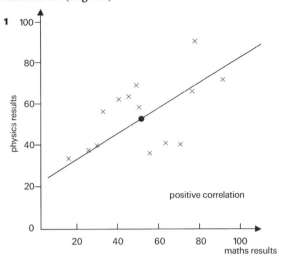
positive correlation

2
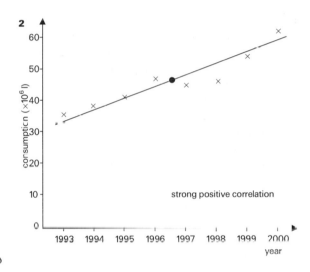
strong positive correlation

3
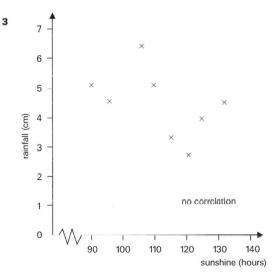
no correlation

4

[Graph showing hours worked per week vs salary (×£1000). Y-axis "hours worker per week" from 0 to 40, X-axis "salary (×£1000)" from 10 to 50. Data points with a line through them, a filled point at approximately (12, 31). Cloud label: "Mean calculated after discarding the outlier" with arrows. Label "positive correlation".]

5

[Graph showing ice-cream sold (×100 l) vs mean temperature (°C). Y-axis "ice-cream sold (×100 l)" from 0 to 25, X-axis "mean temperature (°C)" from 8 to 18. Data points with line through them, filled point at approximately (13, 14.5). Label "positive correlation".]

6

[Graph showing age (years) vs reaction time (×10⁻³ s). Y-axis "age (years)" from 0 to 60, X-axis "reaction time (×10⁻³ s)" from 140 to 210. Data points with line through them, filled point at approximately (180, 42). Label "positive correlation".]

❓ **(Page 96)**

1 Both: independent, random

2 Year: independent, non-random, controlled
Consumption: possibly dependent, random

3 Both: independent, random

4 Salary: possibly dependent, random
Hours worked: independent, random

5 Temperature: independent, random
Ice-cream sold: possibly dependent, random

6 Reaction time: probably dependent, random
Age: independent, random

Exercise 4B (Page 102)

1 **(a)** -0.8 **(b)** 0 **(c)** 0.8

2 $-7, -0.96$

3 $11, 0.704$

4 $-3.2, -0.924$

5 $-8.875, -0.635$

6 $-54.5, -0.128$

Exercise 4C (Page 109)

1 **(i)** 0.913 **(ii)** $H_0: \rho = 0$, $H_1: \rho > 0$ **(iii)** Accept H_1

2 **(i)** 0.380
(ii) $H_0: \rho = 0$ (Jamila), $H_1: \rho > 0$ (coach)
(iii) Accept H_0; there is not enough evidence to reject H_0. r needs to be > 0.4973 to reject H_0 at the 5% significance level.

3 **(i)** 0.715 **(ii)** $H_0: \rho = 0$, $H_1: \rho > 0$
(iii) Accept H_1; it seems that performance in high jump and long jump have positive correlation.

4 **(i)** 0.850 **(ii)** $H_0: \rho = 0$, $H_1: \rho > 0$
(iii) Accept H_1; Yes

5 **(i)** 0.946 **(ii)** $H_0: \rho = 0$, $H_1: \rho > 0$
(iii) Accept H_1; there is very strong positive correlation.

6 $H_0: \rho = 0$, $H_1: \rho > 0$; $r = 0.901$; Andrew

7 **(i)** $H_0: \rho = 0$, $H_1: \rho \neq 0$, 5% sig. level
(ii) 0.491, accept H_1
(iii)

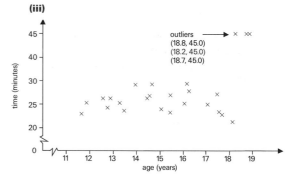

[Scatter graph of time (minutes) vs age (years). Y-axis "time (minutes)" from 0 to 45, X-axis "age (years)" from 11 to 19. Label "outliers → × ××" with coordinates (18.8, 45.0), (18.2, 45.0), (18.7, 45.0).]

Outliers: (18.8, 45), (18.2, 45), (18.7, 45), it seems as though these girls stopped for a rest.

(iv) The scatter diagram should have been drawn first and the outliers investigated before calculating the product moment correlation coefficient. With the three outliers removed, $r = -0.1612$, accept H_0.

8 (i) 0.807   (ii) H_0: $\rho = 0$, H_1: $\rho > 0$ (iii) Accept H_1

(iv)

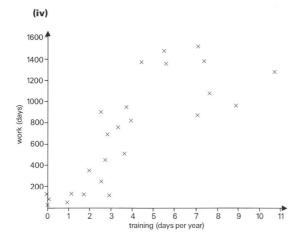

(v) Giving more training to employees does tend to keep staff with the company.

9 (i) H_0: $\rho = 0$, H_1: $\rho < 0$; $r = -0.854$, accept H_1
(ii) Correlation does not imply causation. Perhaps Charlotte ought to gather some data herself to highlight the hazards of drinking alcohol, e.g. wine consumption/liver disease.

10 $r = 0.59$. Diagram suggests moderate positive correlation which is confirmed by the fairly high positive value of r. $r = -0.145$. Discarding high and low values of x seems to produce an uncorrelated set.

11 (i) 0.032   (i) -0.862 (iii) 0.0648   (iv) 0.988
(v) Cases (i) and (iii) indicate no correlation, (ii) strong negative correlation, and (iv) strong positive correlation, indicating a four-quarter cyclic variation in the data.

12 Many other variables are likely to influence the exchange rate of a nation. It is not realistic to affirm any correlation in the variables in this question.

Exercise 4D (Page 120)

1 0.6

2 0.576

3

Brand	A	B	C	D	E	F	G
X's mark	4	3	1	6	7	5	2
Y's mark	5	2	3	4	6	7	1

0.714

4 (i) -0.821 (ii) Yes; judges not in agreement.

5 -0.0875; Use test C and possibly one other test.

6 (i) Y and Z
(ii) H_0: $\rho = 0$, H_1: $\rho > 0$, accept H_1 at 5% significance level.

7 (i) 0.382
(ii) H_0: $\rho = 0$, H_1: $\rho > 0$, judges not in agreement.

8 (i) 0.766, -0.143
(ii) H_0: $\rho = 0$, H_1: $\rho > 0$, accept H_1.
(iii) Product moment correlation coefficient is more suitable here because it takes into account the magnitude of the variables.

9 (i) 0.636. Positive sign indicates possible positive correlation.
(ii) H_0: $\rho = 0$, H_1: $\rho \neq 0$, accept H_1; There is some association between time taken and quality of work.

10 (i) 0.680
(ii) H_0: $\rho = 0$, H_1: $\rho > 0$, $0.680 > 0.6694$ so accept H_1.
(iii) 0.214
(iv) The two correlation coefficients measure different quantities. The product moment correlation coefficient measures linear correlation using the actual data values. Spearman's coefficient measures rank correlation using the data ranks. Mr Smith ought to have used Spearman's coefficient.

11 (i) 0.636   (ii) H_0: $\rho = 0$, H_1: $\rho > 0$, accept H_0
(iii) It is more appropriate to use the product moment correlation coefficient since it utilises the actual data values.

12 (i) 0.881   (ii) 0.888   (iii) 0.622
There is significant correlation at the 5% level in all three cases. There are outliers in (i) contraceptives, (ii) contraceptives and (iii) nuclear power.

Exercise 4E (Page 128)

1 $19x + 50y = 1615$, 27.7

2 $1.446x + y = 114.38$, 53.7

3 (i)

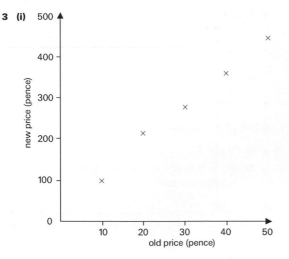

(ii) $y = 8.45x + 27.5$

(iii) Value of this stamp not compatible with earlier part of question.

4 (i)

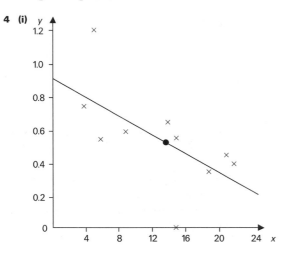

(ii) $0.0289x + y = 0.925$ **(iii)** 0.434

(iv) Stem density is outside domain of validity. If $x > 32$ model predicts that y is negative.

5

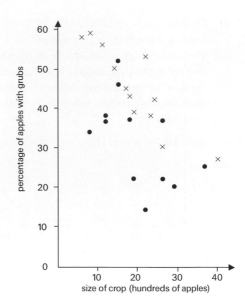

$1.013x + y = 64.25$, 44%. Second treatment seems to produce fewer grubs but linear correlation is not as strong as before.

6 (i)

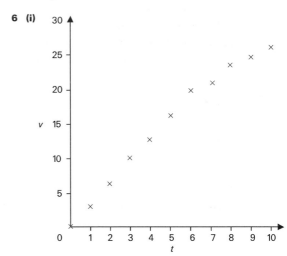

(ii) $v = 2.68t + 1.54$ **(iii)** 0.988

(iv) Modelling data by a single straight line assumes that the correlation is linear. However, looking at the scatter diagram there is a possibility that r is proportional to a power of t thus making the correlation non-linear.

7 (i)

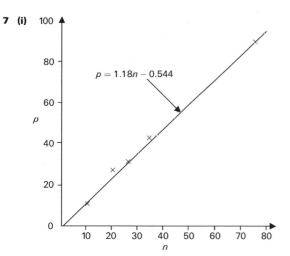

$p = 1.18n - 0.544$

(ii) $p = 1.18n - 0.544$; £17

(iii) 0.998; Both correlation coefficient and scatter diagram suggest almost perfect positive correlation.

8 (i) -180.4 **(ii)** -0.926

(iii) $0.752x + y = 74.97$. Product moment correlation coefficient is preferable because it is a standardised measure of correlation whereas covariance is not.

9 (i)

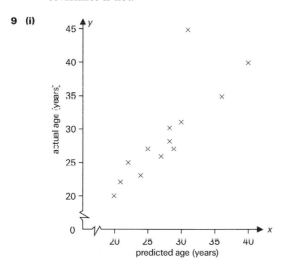

(ii) $y = 1.031x + 0.528$

(iii) Predictions seem very accurate despite the data for person G. A formal hypothesis test is the next step.

10 (i)

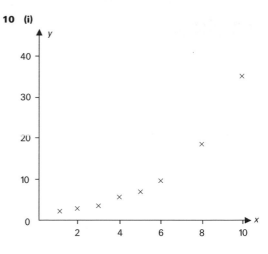

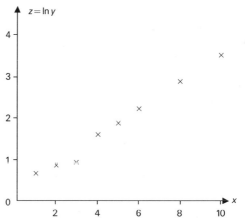

(ii) The line of regression of y on x is not very accurate because the correlation is not linear. The line of regression of z on x is much more accurate.

(iii) 12.8

11 Regression line of y on x: $y = 0.78x + 24.0$ or approx. $y = x + \frac{1}{4}(100 - x)$, thus $\approx \frac{1}{4}$.

12 $m = 5.34 + 0.41d$.

Calculate Pearson's product moment correlation coefficient and then conduct a hypothesis test to decide whether there is any correlation. $r = 0.607$.

$H_0: \rho = 0$;

$H_1: \rho > 0$. Conclusion: reject H_0.

Index

binomial distribution 30, 36–39
  approximating the binomial
    distribution 36–39
  derivation of mean 134
  derivation of variance 135
bivariate data 89–133
  line of best fit 91, 94–95
    extrapolation of line of best fit
      109
  product moment correlation 96

central limit theorem 84
continuity correction 76
correlation 92, 94
  interpreting correlation 108–109
  linear correlation 92, 94, 125
  negative correlation 92, 97
  non-linear correlation 108–109,
    119
  positive correlation 92, 97
correlation coefficient 100, 103–104
  hypothesis test 103–104
covariance, see sample covariance
cumulative distribution function 5

degrees of freedom 107–108
discrete random variables 1–35
  conditions 4
  notation 4

exhaustive outcomes 4
expectation (or mean) 10–16
  definition 12
  expectation algebra 15–16
  expectation of a function of X 14

geometric distribution 31

hypothesis test 103–104, 118–120
  correlation coefficient 103–104
  rank correlation 118–120

least squares regression line 125–127
  derivation 139–141
linear correlation, see correlation
line of best fit 91, 94–95, 109

mathematical model 2
mean, see expectation

non-linear correlation, see
  correlation
non-parametric test 120
normal distribution 59–88
  approximating the binomial
    distribution 77
  approximating the Poisson
    distribution 79
  inverse normal distribution tables
    68
  modelling discrete situations
    75–80
  normal curve 60, 65–66
  normal distribution tables 61–62
  normal probability graph paper
    85–87
  standardised normal distribution
    62, 66
  z-value 61–62

parameters 12
Pearson, Karl 101
Pearson's product moment
  correlation coefficient 99–100,
    119
Poisson distribution 35–58
  as a distribution in its own
    right 41
  as an approximation to binomial
    distribution 39–41
  cumulative Poisson probability
    tables 43
  definition 39
  derivation of mean 135
  derivation of variance 136
  notation 42
  population parameter 39, 44
  recurrence relations 42–43
  sum of two or more Poisson
    distributions 51–52, 58
    derivation of formula for sum of
      two Poisson distributions
      137
  variance 44
Poisson, Simeon 57
probability density function 5
probability distribution 4–6

product moment correlation 96–107
  Pearson's product moment
    correlation 99–100, 119

rank correlation 115–120
  equivalence of Spearman's and
    Pearson's correlation
    coefficients 137–138
  hypothesis test 118–120
  Spearman's coefficient of rank
    correlation 117, 120
  tied ranks 119
  when to use rank correlation
    119–120
recurrence relations 42–43
regression line, see least squares
  regression line
residual 126

sample covariance 98–99
scatter diagrams 90–94
skewness 87
  negative skew 87
  positive skew 87
Spearman, Charles 119
Spearman's coefficient of rank
  correlation see rank correlation
standard deviation 19, 60, 61, 62
statistical model 2–3

test statistic 104

uniform distribution 31

variables 91–92
  dependent variables 91, 125
  discrete variables 3
  continuous variables 3
  controlled variables 91–92
  independent variables 91, 125
  non-random variables 91–92
  random variables 4, 91–92
variance 19–24
  variance of a discrete random
    variable 20
  variance of a linear function of
    random variable X 23–24
  variance of a set of numbers 19